Engineering Dynamics

Springer
New York
Berlin
Heidelberg
Barcelona
Hong Kong
London
Milan
Paris
Singapore
Tokyo

Oliver M. O'Reilly

Engineering Dynamics

A Primer

With 61 Illustrations

 Springer

Oliver M. O'Reilly
Department of Mechanical Engineering
6137 Etcheverry Hall
University of California
Berkeley, CA 94720-1740
USA
oreilly@newton.me.berkeley.edu

Library of Congress Cataloging-in-Publication Data
O'Reilly, Oliver M.
    Engineering dynamics: a primer / Oliver M. O'Reilly.
        p.   cm.
    Includes bibliographical references and index.
    ISBN 0-387-95145-8 (softcover: alk. paper)
        1. Dynamics. I. Title.
TA352 .O74 2000
620.1'04—dc21   00-045032

Printed on acid-free paper.

Production managed by A. Orrantia; manufacturing supervised by Jeff Taub.
Camera-ready copy provided by the author.
Printed and bound by Edwards Brothers, Inc., Ann Arbor, MI.
Printed in the United States of America.

9 8 7 6 5 4 3 2 1

ISBN 0-387-95145-8   SPIN 10780652

Springer-Verlag New York Berlin Heidelberg
A member of BertelsmannSpringer Science+Business Media GmbH

Dedicated with Love to Lisa

Preface

Scope, Aims, and Audiences

This primer is intended to provide the theoretical background for the standard undergraduate course in dynamics. This course is usually based on one of the following texts: Bedford and Fowler [6], Beer and Johnston [7], Hibbeler [33], Meriam and Kraige [39], Riley and Sturges [50], and Shames [56], among others. Although most teachers will have certain reservations about these texts, there appears to be a general consensus that the selection of problems each of them presents is an invaluable and essential aid for studying and understanding dynamics.

I myself use Meriam and Kraige [39] when teaching such a course, which is referred to as ME104 at the University of California at Berkeley. However, I have found that the gap between the theory presented in the aforementioned texts and the problems I wished my students to solve was too large. As a result, I prepared my own set of notes on the relevant theory, and I used Meriam and Kraige [39] as a problem and homework resource. This primer grew out of these notes. Its content was also heavily influenced by three other courses that I teach: one on rigid body dynamics, one on Lagrangian mechanics, and another on Hamiltonian mechanics.[1] Because I use the primer as a supplement, I have only included a set of brief exercises at the end of each chapter. Further, dimensions of physical quantities and

---

[1] These courses are referred to as ME170, ME175, and ME275 in the University of California at Berkeley course catalog.

numerical calculations are not emphasized in the primer because I have found that most students do not have serious problems with these matters.

This primer is intended for three audiences: students taking an undergraduate engineering dynamics course, graduate students needing a refresher in such a course, and teachers of such a course. For the students, I hope that this primer succeeds in providing them with a succinct account of the theory needed for the course, an exploration of the limitations of such a course, and a message that the subject at hand can be mastered with understanding and not rote memorization of formulae. For all of these audiences, an appendix provides the notational and presentational correspondences between the chapters in this primer and the aforementioned texts. In addition, each chapter is accompanied by a summary section.

I have noticed an increased emphasis on "practical" problems in engineering dynamics texts. Although such an emphasis has its merits, I think that the most valuable part of an education is the evolution and maturation of the student's thinking abilities and thought processes. With this in mind, I consider the development of the student's analytical skills to be paramount. This primer reflects my philosophy in this respect.

The material in this primer is not new. I have merely reorganized some classical thoughts and theories on the subject in a manner which suits an undergraduate engineering dynamics course. My sources are contained in the references section at the end of this primer. Apart from the engineering texts listed above, the works of Beatty [5], Casey [12, 14], and Synge and Griffith [64] had a significant influence on my exposition.

I have also included some historical references and comments in this primer in the hopes that some students may be interested in reading the original work. Most of the historical information in the primer was obtained from Scribner's *Dictionary of Scientific Biography*. I heartily recommend reading the biographies of Euler, Kepler, Leibniz, and others contained in this wonderful resource.

Finally, have tried where ever possible to outline the limitations of what is expected from a student for two reasons. First, some students will decide to extend their knowledge beyond these limitations, and, second, it gives a motivation to the types of questions asked of the student.

Acknowledgments

My perspective on dynamics has been heavily influenced by both the continuum mechanics and dynamics communities. I mention in particular the writings and viewpoints of Jim Casey, Jim Flavin, Phil Holmes, Paul M. Naghdi, Ronald Rivlin, and Clifford Truesdell. I owe a large debt of gratitude to Jim Casey both for showing me the intimate relationship between continuum mechanics and dynamics, and for supporting my teaching here at Berkeley.

The typing of this primer using LATEXwould not have been possible without the assistance of Bonnie Korpi, Laura Cantú, and Linda Witnov. Laura helped with the typing of Chapters 7, 8, 9, and 10. Linda did the vast majority of the work on the remaining chapters. Her patience and cheerful nature in dealing with the numerous revisions and reorganizations was a blessing for me. David Kramer was a copyreader for the primer, and he provided valuable corrections to the final version of the primer. The publication of this primer was made possible by the support of Achi Dosanjh at Springer-Verlag. Achi also organized two sets of helpful reviews. Several constructive criticisms made by the anonymous reviewers have been incorporated, and I would like to take this opportunity to thank them.

Many of my former students have contributed directly and indirectly to this primer. In particular, Tony Urry read through an earlier draft and gave numerous insightful comments on the presentation. I have also benefited from numerous conversations with my former graduate students Tom Nordenholz, Jeffrey Turcotte, and Peter Varadi. As I mentioned earlier, this primer arose from my lecture notes for ME104. My interactions with the former students in this course have left an indelible impression on this primer.

Finally, I would like to thank my wife, Lisa, my parents, Anne and Jackie, and my siblings, Séamus and Sibéal, for their support and encouragement.

Oliver M. O'Reilly   Berkeley

Contents

1
Elementary Particle Dynamics

Here, we cover the basics on kinematics and kinetics of particles and discuss three ubiquitous examples. We conclude with a discussion of Euler's first law (Newton's second law, or the balance of linear momentum). Much of the material in this chapter will subsequently be repeated.

1.1 Kinematics of a Particle

Consider a particle moving in a three-dimensional space $\mathcal{E}^3$. The position vector $\mathbf{r}$ of the particle relative to a fixed origin as a function of time is denoted by the function $\mathbf{r}(t)$, i.e., given a time t the location of the particle is determined by the value $\mathbf{r} = \mathbf{r}(t)$ (see Figure 1.1). Varying t, $\mathbf{r}(t)$ defines the motion and the path $\mathcal{C}$ of the particle. This path in many cases coincides with a specific curve, for example, a particle moving on a circular ring or a particle in motion on a circular helix. Otherwise, the particle is either free or in motion on a surface.

The (absolute) velocity vector $\mathbf{v}$ of the particle can be determined by differentiating $\mathbf{r}(t)$ with respect to time t:

$$\mathbf{v} = \mathbf{v}(t) = \frac{d\mathbf{r}}{dt} = \lim_{\triangle t \to 0} \frac{\mathbf{r}(t + \Delta t) - \mathbf{r}(t)}{\Delta t} .$$

The speed of the particle is given by the magnitude of the velocity vector $\|\mathbf{v}\|$. We often denote the time derivative of a function by a superposed dot,

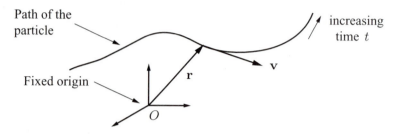

FIGURE 1.1. Some kinematical quantities pertaining to a particle and its motion

for example, $\mathbf{v} = \dot{\mathbf{r}}$. The (absolute) acceleration vector $\mathbf{a}$ of the particle is determined by differentiating the (absolute) velocity vector with respect to time:

$$\mathbf{a} = \mathbf{a}(t) = \frac{d\mathbf{v}}{dt} = \lim_{\Delta t \to 0} \frac{\mathbf{v}(t + \Delta t) - \mathbf{v}(t)}{\Delta t}.$$

To calculate the distance traveled by a particle along its path, it is convenient to introduce a parameter s: the arc-length parameter. This parameter is defined by

$$\frac{ds}{dt} = \|\mathbf{v}\|.$$

Clearly, $ds/dt \geq 0$ is the speed of the particle. Integrating this relationship one finds that

$$s(t) - s_0 = \int_{t_0}^{t} \frac{ds}{dt}(\tau)d\tau = \int_{t_0}^{t} \sqrt{\mathbf{v}(\tau) \cdot \mathbf{v}(\tau)}\, d\tau.$$

It should be noted that $s(t) - s_0$ is the distance traveled by the particle along its path $\mathcal{C}$ during the time interval $t - t_0$. Also, $s(t_0) = s_0$, where t_0 and s_0 are initial conditions. You should also notice that we use a dummy variable τ when performing the integration.[1]

Often, instead of using t to parametrize the motion of the particle, one uses the arc-length parameter s. We will see several examples of this parametrization in Chapter 3. The motion will, in general, be a different function of s than it is of t. To distinguish these functions, we denote the motion as a function of t by $\mathbf{r}(t)$, while the motion as a function of s is denoted by $\hat{\mathbf{r}}(s)$. Provided $\dot{s}$ is never zero, these functions provide the same value of $\mathbf{r}$: $\mathbf{r}(t) = \hat{\mathbf{r}}(s(t))$.[2]

---

[1]It is necessary to use a dummy variable τ as opposed to the variable t when evaluating this integral because we are integrating the magnitude of the velocity as τ varies between t_0 and t. If we take the derivative with respect to t of the integral, then, using the fundamental theorem of calculus, we would find, as expected, that $\dot{s}(t) = \|\mathbf{v}(t)\|$. Had we not used the dummy variable τ but rather t to perform the integration, then the derivative of the resulting integral with respect to t would not yield $\dot{s}(t) = \|\mathbf{v}(t)\|$.

[2]Here, we are invoking the inverse function theorem of calculus. If $\dot{s}$ were zero, then the particle would be stationary, s would be constant, but time would continue increasing, so there would not be a one-to-one correspondence between s and t.

With the above proviso in mind, one has the following relations:

$$
\begin{aligned}
s &= s(t) = s_0 + \int_{t_0}^{t} \sqrt{\mathbf{v}(\tau) \cdot \mathbf{v}(\tau)} \, d\tau \,, \\
\mathbf{r} &= \mathbf{r}(t) = \hat{\mathbf{r}}(s(t)) \,, \\
\mathbf{v} &= \frac{d\mathbf{r}(t)}{dt} = \frac{d\mathbf{r}}{ds} \frac{ds}{dt} \,, \\
\mathbf{a} &= \frac{d\mathbf{v}(t)}{dt} = \frac{d^2\mathbf{r}}{ds^2} \left(\frac{ds}{dt} \right)^2 + \frac{d\mathbf{r}}{ds} \frac{d^2 s}{dt^2} \,.
\end{aligned}
$$

At this stage, we haven't used a particular coordinate system, so all of the previous results are valid for any coordinate system. For most of the remainder of this primer we will use three different sets of orthonormal bases: Cartesian $\{\mathbf{E}_x, \mathbf{E}_y, \mathbf{E}_z\}$, cylindrical polar $\{\mathbf{e}_r, \mathbf{e}_\theta, \mathbf{E}_z\}$, and the Serret-Frenet triad $\{\mathbf{e}_t, \mathbf{e}_n, \mathbf{e}_b\}$. Which set one uses depends on the problem of interest. Knowing which one to select is an art, and to acquire such experience is very important.[3]

1.2   A Circular Motion

We now elucidate the preceding developments with a simple example. Suppose the position vector of a particle has the representation

$$
\mathbf{r} = \mathbf{r}(t) = a(\cos(\omega t)\mathbf{E}_x + \sin(\omega t)\mathbf{E}_y) \,,
$$

where ω is a positive constant and a is greater than zero. In Figure 1.2, the path of the particle is shown. You should notice that the path of the particle is a circle of radius a that is traversed in a counterclockwise direction.[4] The question we seek to answer here is, what are $\mathbf{v}(t)$, $\mathbf{a}(t)$, $s(t)$, $t(s)$, $\hat{\mathbf{v}}(s)$, and $\hat{\mathbf{a}}(s)$?

First, let's calculate $\mathbf{v}(t)$ and $\mathbf{a}(t)$:

$$
\begin{aligned}
\mathbf{v} &= \dot{\mathbf{r}} = a\omega(-\sin(\omega t)\mathbf{E}_x + \cos(\omega t)\mathbf{E}_y) \,, \\
\mathbf{a} &= \ddot{\mathbf{r}} = \dot{\mathbf{v}} = -a\omega^2(\cos(\omega t)\mathbf{E}_x + \sin(\omega t)\mathbf{E}_y) = -\omega^2 \mathbf{r} \,.
\end{aligned}
$$

Next,

$$
s(t) - s_0 = \int_{t_0}^{t} \sqrt{\mathbf{v}(\tau) \cdot \mathbf{v}(\tau)} \, d\tau = \int_{t_0}^{t} \sqrt{\omega^2 a^2} d\tau = a\omega \, (t - t_0) \,.
$$

---

[3]In other words, the more problems one examines, the better.

[4]Later on, hopefully sooner rather than later, you should revisit this problem using the cylindrical polar coordinates, $r = a$ and $\theta = \omega t$, and establish the forthcoming results using cylindrical polar basis vectors.

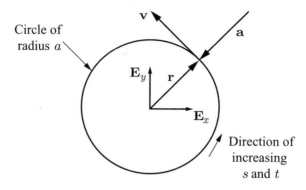

FIGURE 1.2. Path of a particle in a circular motion

Hence,

$$t(s) = t_0 + \frac{1}{a\omega}\left(s - s_0\right).$$

The last formula allows us to write down the following results with a minimum of effort using $\mathbf{v}(t)$ and $\mathbf{a}(t)$:

$$\mathbf{v} = \hat{\mathbf{v}}(s) = a\omega\left(-\sin\left(\frac{s - s_0}{\omega a} + t_0\right)\mathbf{E}_x + \cos\left(\frac{s - s_0}{\omega a} + t_0\right)\mathbf{E}_y\right),$$

$$\mathbf{a} = \hat{\mathbf{a}}(s) = -a\omega^2\left(\cos\left(\frac{s - s_0}{\omega a} + t_0\right)\mathbf{E}_x + \sin\left(\frac{s - s_0}{\omega a} + t_0\right)\mathbf{E}_y\right)$$

$$= -\omega^2\hat{\mathbf{r}}(s).$$

Alternatively, one could use the expression for $t(s)$ to determine $\mathbf{r}$ as a function of s, and then differentiate with respect to t to obtain the desired functions.

1.3   Rectilinear Motions

In this section, we consider the motion of a particle along a straight line. We take $\mathbf{E}_x$ to be parallel to this line and the vector $\mathbf{c}$ to be a constant. Then,

$$\mathbf{r} = \mathbf{r}(t) = x(t)\mathbf{E}_x + \mathbf{c},$$

$$\mathbf{v} = \mathbf{v}(t) = \frac{dx}{dt}\mathbf{E}_x = v(t)\mathbf{E}_x,$$

$$\mathbf{a} = \mathbf{a}(t) = \frac{d^2x}{dt^2}\mathbf{E}_x = a(t)\mathbf{E}_x.$$

It is important to note that

$$\frac{ds}{dt} = \left|\frac{dx}{dt}\right|,$$

so unless $\dot{x} > 0$ or $\dot{x} < 0$, x and s cannot be easily interchanged.

The material that follows should be familiar to you from other courses. Further, graphical interpretations of the forthcoming results are readily available. Essentially, they involve relating t, x, v, and a. There are three cases to consider.

1.3.1 Given Acceleration as a Function of Time

Suppose one knows $a(t)$, then $v(t)$ and $x(t)$ can be determined by integrating $a(t)$:

$$v(t) = v(t_0) + \int_{t_0}^{t} a(u)du \,,$$

$$x(t) = x(t_0) + \int_{t_0}^{t} v(\tau)d\tau = x(t_0) + \int_{t_0}^{t} \left(v(t_0) + \int_{t_0}^{\tau} a(u)du \right) d\tau \,.$$

1.3.2 Given Acceleration as a Function of Speed

The next case to consider is when $a = \bar{a}(v)$ and one is asked to calculate $\bar{x}(v)$ and $\bar{t}(v)$. The former task is achieved by noting that $dt = \frac{dv}{a}$, while the latter task is achieved by using the identities

$$a = \frac{dv}{dt} = \left(\frac{dv}{dx} \right) \left(\frac{dx}{dt} \right) = v\frac{dv}{dx} \,.$$

In summary,

$$\bar{t}(v) = \bar{t}(v_0) + \int_{v_0}^{v} \frac{1}{\bar{a}(u)} du \,,$$

$$\bar{x}(v) = \bar{x}(v_0) + \int_{v_0}^{v} \frac{u}{\bar{a}(u)} du \,.$$

We remark that the identity $a = v\frac{dv}{dx}$ is very useful.

1.3.3 Given Acceleration as a Function of Placement

The last case to consider is when $a = \hat{a}(x)$ is known, and one seeks $\hat{v}(x)$ and $\hat{t}(x)$. Again, the former result is calculated using the identity $a = v\frac{dv}{dx}$, and the latter is calculated using the identity $dt = \frac{dx}{v}$:

$$\hat{v}^2(x) = \hat{v}^2(x_0) + 2\int_{x_0}^{x} \hat{a}(u)du \,,$$

$$\hat{t}(x) = \hat{t}(x_0) + \int_{x_0}^{x} \frac{du}{\hat{v}(u)} \,.$$

1.4   Kinetics of a Particle

Consider a particle of constant mass m. Let $\mathbf{F}$ denote the resultant external force acting on the particle, and let $\mathbf{G} = m\mathbf{v}$ be the linear momentum of the particle. Euler's first law[5] (also known as Newton's second law[6] or the balance of linear momentum) postulates that

$$\mathbf{F} = \frac{d\mathbf{G}}{dt} = m\mathbf{a}.$$

It is crucial to note that $\mathbf{a}$ is the absolute acceleration vector of the particle. We shall shortly write this equation with respect to several sets of basis vectors. For instance, with respect to a (right-handed) Cartesian basis, the vector equation $\mathbf{F} = m\mathbf{a}$ is equivalent to 3 scalar equations:

$$
\begin{aligned}
F_x &= ma_x = m\frac{d^2x}{dt^2}, \\
F_y &= ma_y = m\frac{d^2y}{dt^2}, \\
F_z &= ma_z = m\frac{d^2z}{dt^2},
\end{aligned}
$$

where

$$\mathbf{F} = F_x\mathbf{E}_x + F_y\mathbf{E}_y + F_z\mathbf{E}_z, \quad \mathbf{a} = a_x\mathbf{E}_x + a_y\mathbf{E}_y + a_z\mathbf{E}_z.$$

1.4.1   Action and Reaction

When dealing with the forces of interaction between particles, a particle and a rigid body, and rigid bodies, we shall also invoke Newton's third law: "For every action there is an equal and opposite reaction." For example, consider a particle moving on a surface. From this law, the force exerted by the surface on the particle is equal in magnitude and opposite in direction to the force exerted by the particle on the surface.

1.4.2   The Four Steps

There are four steps to solving problems using $\mathbf{F} = m\mathbf{a}$:

---

[5]Leonhard Euler (1707–1783) made enormous contributions to mechanics and mathematics. We follow C. Truesdell (see Essays II and V in [65]) in crediting $\mathbf{F} = m\mathbf{a}$ to Euler. As noted by Truesdell, these differential equations can be seen on pages 101–105 of a 1749 paper by Euler [22]. Truesdell's essays also contain copies of certain parts of a related seminal paper [23] by Euler that was published in 1752.

[6]Isaac Newton (1642–1727) wrote his second law in Volume 1 of his famous *Principia* in 1687 as follows: *The change of motion is proportional to the motive force impressed; and is made in the direction of the right line in which that force is impressed.* (Cf. page 13 of [43].)

1. Pick an origin and a coordinate system, and then establish expressions for $\mathbf{r}, \mathbf{v}$, and $\mathbf{a}$.

2. Draw a free-body diagram.

3. Write out $\mathbf{F} = m\mathbf{a}$.

4. Perform the analysis.

These four steps will guide you through most problems. We will amend them later on, in an obvious way, when dealing with rigid bodies. If you follow them they will help you with homeworks and exams – although I realize that for many readers they will seem to be overkill and an enormous stunting of their creativity.

One important point concerns the free-body diagram. This is a graphical summary of the external forces acting on the particle. It does not include any accelerations. Here, in contrast to some other treatments, it is used only as an easy visual check on one's work.

1.5   A Particle Under the Influence of Gravity

Consider a particle of mass m that is launched with an initial velocity $\mathbf{v}_0$ at $t = 0$. At this instant, $\mathbf{r} = \mathbf{r}_0$. During the subsequent motion of the particle it is under the influence of a vertical gravitational force $-mg\mathbf{E}_y$. In S.I. units, g is approximately 9.81 meters per second per second. One is asked to determine the path $\mathbf{r}(t)$ of the particle.

The example of interest is a standard projectile problem. It also provides a model for the motion of the center of mass of many falling bodies where the influence of drag forces is ignored. For example, it is a model for a vehicle falling through the air. To determine the motion of the particle predicted by this model, we will follow the four aforementioned steps.

1.5.1   *Kinematics*

For this problem it is convenient to use a Cartesian coordinate system. One then has the representations

$$\mathbf{r} = x\mathbf{E}_x + y\mathbf{E}_y + z\mathbf{E}_z , \quad \mathbf{a} = \ddot{x}\mathbf{E}_x + \ddot{y}\mathbf{E}_y + \ddot{z}\mathbf{E}_z .$$

1.5.2   *Forces*

The sole force acting on the particle is gravity, so $\mathbf{F} = -mg\mathbf{E}_y$ and the free-body diagram is trivial. It is shown in Figure 1.3.

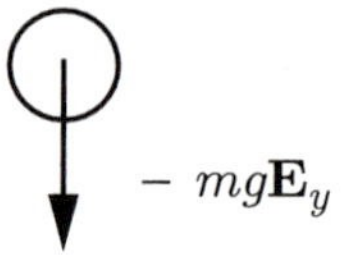

$$- mg\mathbf{E}_y$$

FIGURE 1.3. Free-body diagram of a particle in a gravitational field

1.5.3 Balance Law

From $\mathbf{F} = m\mathbf{a}$, we obtain 3 second-order ordinary differential equations:

$$
\begin{aligned}
m\ddot{x} &= 0, \\
m\ddot{y} &= -mg, \\
m\ddot{z} &= 0.
\end{aligned}
$$

1.5.4 Analysis

The final step in solving the problem involves finding the solution to the previous differential equations that satisfies the given initial conditions:

$$
\begin{aligned}
\mathbf{r}_0 &= \mathbf{r}(t=0) = x_0\mathbf{E}_x + y_0\mathbf{E}_y + z_0\mathbf{E}_z, \\
\mathbf{v}_0 &= \mathbf{v}(t=0) = \dot{x}_0\mathbf{E}_x + \dot{y}_0\mathbf{E}_y + \dot{z}_0\mathbf{E}_z.
\end{aligned}
$$

The differential equations in question have simple solutions:

$$
\begin{aligned}
x(t) &= \dot{x}_0 t + x_0, \\
y(t) &= -\frac{1}{2}gt^2 + \dot{y}_0 t + y_0, \\
z(t) &= \dot{z}_0 t + z_0.
\end{aligned}
$$

Hence, the motion of the particle can be written in a compact form:

$$
\mathbf{r}(t) = \mathbf{r}_0 + \mathbf{v}_0 t - \frac{1}{2}gt^2\mathbf{E}_y.
$$

By specifying particular sets of initial conditions, the results for the motion of the particle apply to numerous special cases.

1.6 Summary

In this chapter, several definitions of kinematical quantities pertaining to a single particle were presented. In particular, the position vector $\mathbf{r}$ relative to a fixed origin was defined. This vector defines the path of the particle. Furthermore, the velocity $\mathbf{v}$ and acceleration $\mathbf{a}$ vectors were defined:

$$
\mathbf{v} = \mathbf{v}(t) = \frac{d\mathbf{r}}{dt}, \quad \mathbf{a} = \mathbf{a}(t) = \frac{d^2\mathbf{r}}{dt^2}.
$$

These vectors can also be defined as functions of the arc–length parameter s: $\mathbf{v} = \hat{\mathbf{v}}(s)$ and $\mathbf{a} = \hat{\mathbf{a}}(s)$. Here, s is defined by integrating the differential equation $\dot{s} = \|\mathbf{v}\|$. The parameter s can be used to determine the distance traveled by the particle along its path. Using the chain rule, the following results were established

$$\mathbf{v} = \dot{s}\frac{d\mathbf{r}}{ds}, \quad \mathbf{a} = \dot{s}^2\frac{d^2\mathbf{r}}{ds^2} + \ddot{s}\frac{d\mathbf{r}}{ds}.$$

Two special cases of the aforementioned results were discussed in Sections 1.2 and 1.3. First, in Section 1.2, the kinematics of a particle moving in a circular path was discussed. Then, in Section 1.3, the corresponding quantities pertaining to rectilinear motion were presented.

The balance of linear momentum $\mathbf{F} = m\mathbf{a}$ was then introduced. This law relates the motion of the particle to the resultant force $\mathbf{F}$ acting on the particle. In Cartesian coordinates, it can be written as the following three scalar equations:

$$\begin{aligned} F_x &= ma_x = m\ddot{x}\,, \\ F_y &= ma_y = m\ddot{y}\,, \\ F_z &= ma_z = m\ddot{z}\,, \end{aligned}$$

where

$$\mathbf{F} = F_x\mathbf{E}_x + F_y\mathbf{E}_y + F_z\mathbf{E}_z\,, \quad \mathbf{a} = a_x\mathbf{E}_x + a_y\mathbf{E}_y + a_z\mathbf{E}_z\,.$$

In order to develop a helpful problem solving methodology, a series of four steps were introduced. These steps are designed to provide a systematic framework to help guide you through problems. To illustrate the steps, the classic projectile problem was discussed in Section 1.5.

1.7   Exercises

The following short exercises are intended to assist you in reviewing Chapter 1.

1.1 Why are the time derivatives of $\mathbf{E}_x$, $\mathbf{E}_y$, and $\mathbf{E}_z$ zero?

1.2 Suppose that you are given a vector as a function of s: $\mathbf{f}(s)$. Why do you need to know how s depends on time t in order to determine the derivative of $\mathbf{f}$ as a function of time?

1.3 Consider a particle of mass m which lies at rest on a horizontal surface. A vertical gravitational force $-mg\mathbf{E}_y$ acts on the particle. Draw a free–body diagram of the particle. If one follows the four steps, then why is it a mistake to write, for the normal force, $mg\mathbf{E}_y$ instead of $N\mathbf{E}_y$, where N is unknown, in the free–body diagram?

1.4 The motion of a particle is such that its position vector $\mathbf{r}(t) = 10\mathbf{E}_x + 10t\mathbf{E}_y + 5t\mathbf{E}_z$ (meters). Show that the path of the particle is a straight line, and that the particle moves along this line at a constant speed of $\sqrt{125}$ (meters/second). Furthermore, show that the force $\mathbf{F}$ needed to sustain this motion is $\mathbf{0}$.

1.5 The motion of a particle is such that its position vector $\mathbf{r}(t) = 3t\mathbf{E}_x + 4t\mathbf{E}_y + 10\mathbf{E}_z$ (meters). Show that the path of the particle is a straight line, and that the particle moves along this line at a constant speed of 5 (meters/second). Using this information, show that the arc–length parameter s is given by $s(t) = 5(t - t_0) + s_0$. Finally, show that the distance the particle moves 50 meters along its path every 10 seconds.

1.6 The motion of a particle is such that its position vector $\mathbf{r}(t) = 10\cos(n\pi t)\mathbf{E}_x + 10\sin(n\pi t)\mathbf{E}_y$ (meters). Show that the particle is moving on a circle of radius 10 meters and describes a complete circle every $\frac{2}{n}$ seconds. If the particle has a mass of 2 kilograms, then what force $\mathbf{F}$ is needed to sustain this motion?

1.7 To model the free–fall of a ball of mass m, the ball is modeled as a particle of the same mass. Suppose the particle is dropped from the top of a 100 meter high building, then, following the steps discussed in Section 1.5, show that it takes $\sqrt{\frac{200}{9.81}}$ seconds for the ball to reach the ground. Furthermore, show that it will hit the ground at a speed of $\sqrt{1962}$ meters per second.[7]

1.8 A projectile is launched at time $t_0 = 0$ seconds from a location $\mathbf{r}(t_0) = \mathbf{0}$. The initial velocity of the projectile is $\mathbf{v}(t_0) = v_0\cos(\alpha)\mathbf{E}_x + v_0\sin(\alpha)\mathbf{E}_y$. Here, v_0 and α are constants. During its flight, a vertical gravitational force $-mg\mathbf{E}_y$ acts on the projectile. Modeling the projectile as a particle of mass m, show that its path is a parabola: $y(x) = \frac{g}{2v_0^2\cos^2(\alpha)}x^2 + x\tan(\alpha)$. Why is this result not valid when $\alpha = \pm\frac{\pi}{2}$?

---

[7]This speed is equal to 99.09 miles per hour.

2
Particles and Cylindrical Polar Coordinates

Here, we discuss the cylindrical polar coordinate system and how it is used in particle mechanics. This coordinate system and its associated basis vectors $\{\mathbf{e}_r, \mathbf{e}_\theta, \mathbf{E}_z\}$ are vital to understand and practice.

It is a mistake to waste time memorizing formulae here. Instead, focus on understanding the material. You will repeat it countless times, and you will naturally develop the ability to derive the results from scratch.

2.1 The Cylindrical Polar Coordinate System

Consider a Cartesian coordinate system $\{x, y, z\}$ for the three-dimensional space $\mathcal{E}^3$. Using these coordinates, one can define a cylindrical polar coordinate system $\{r, \theta, z\}$:

$$r = \sqrt{x^2 + y^2}\,, \quad \theta = \tan^{-1}\left(\frac{y}{x}\right)\,, \quad z = z\,.$$

Apart from the points $\{x, y, z\} = \{0, 0, z\}$, given r, θ, and z, we can uniquely determine x, y, and z:

$$x = r\cos(\theta)\,, \quad y = r\sin(\theta)\,, \quad z = z\,.$$

Here, θ is taken to be positive in the counterclockwise direction.

If we now consider the position vector $\mathbf{r}$ of a point in this space, we have, as always,

$$\mathbf{r} = x\mathbf{E}_x + y\mathbf{E}_y + z\mathbf{E}_z\,.$$

We can write this position vector using cylindrical polar coordinates by substituting for x and y in terms of r and θ:

$$\mathbf{r} = r\cos(\theta)\mathbf{E}_x + r\sin(\theta)\mathbf{E}_y + z\mathbf{E}_z .$$

Before we use this representation to establish expressions for the velocity and acceleration vectors, it is prudent to pause and define two new vectors.

To simplify subsequent notation, it is convenient to introduce two unit vectors $\mathbf{e}_r$ and $\mathbf{e}_\theta$:

$$
\begin{bmatrix} \mathbf{e}_r \\ \mathbf{e}_\theta \\ \mathbf{E}_z \end{bmatrix}
=
\begin{bmatrix} \cos(\theta) & \sin(\theta) & 0 \\ -\sin(\theta) & \cos(\theta) & 0 \\ 0 & 0 & 1 \end{bmatrix}
\begin{bmatrix} \mathbf{E}_x \\ \mathbf{E}_y \\ \mathbf{E}_z \end{bmatrix} .
$$

Two of these vectors are shown in Figure 2.1.

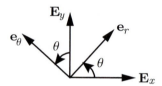

FIGURE 2.1. Two of the cylindrical polar basis vectors

Note that $\{\mathbf{e}_r, \mathbf{e}_\theta, \mathbf{E}_z\}$ are orthonormal, and form a right-handed basis[1] for $\mathcal{E}^3$. You should also be able to see that

$$
\begin{bmatrix} \mathbf{E}_x \\ \mathbf{E}_y \\ \mathbf{E}_z \end{bmatrix}
=
\begin{bmatrix} \cos(\theta) & -\sin(\theta) & 0 \\ \sin(\theta) & \cos(\theta) & 0 \\ 0 & 0 & 1 \end{bmatrix}
\begin{bmatrix} \mathbf{e}_r \\ \mathbf{e}_\theta \\ \mathbf{E}_z \end{bmatrix} .
$$

Since $\mathbf{e}_r = \mathbf{e}_r(\theta)$ and $\mathbf{e}_\theta = \mathbf{e}_\theta(\theta)$, these vectors change as θ changes:

$$
\frac{d\mathbf{e}_r}{d\theta} = -\sin(\theta)\mathbf{E}_x + \cos(\theta)\mathbf{E}_y = \mathbf{e}_\theta ,
$$

$$
\frac{d\mathbf{e}_\theta}{d\theta} = -\cos(\theta)\mathbf{E}_x - \sin(\theta)\mathbf{E}_y = -\mathbf{e}_r .
$$

It is crucial to note that θ is measured positive in the counterclockwise direction.

Returning to the position vector $\mathbf{r}$, it follows that

$$\mathbf{r} = r\mathbf{e}_r + z\mathbf{E}_z = x\mathbf{E}_x + y\mathbf{E}_y + z\mathbf{E}_z .$$

Furthermore, since $\{\mathbf{e}_r, \mathbf{e}_\theta, \mathbf{E}_z\}$ is a basis, we then have, for any vector $\mathbf{b}$, that

$$\mathbf{b} = b_r\mathbf{e}_r + b_\theta\mathbf{e}_\theta + b_z\mathbf{E}_z = b_x\mathbf{E}_x + b_y\mathbf{E}_y + b_z\mathbf{E}_z .$$

It should be clear that $b_r = \mathbf{b} \cdot \mathbf{e}_r$, $b_\theta = \mathbf{b} \cdot \mathbf{e}_\theta$, and $b_z = \mathbf{b} \cdot \mathbf{E}_z$.

---

[1] Details on these results are discussed in Appendix A.

2.2   Velocity and Acceleration Vectors

Consider a particle moving in space: $\mathbf{r} = \mathbf{r}(t)$. We have that

$$\mathbf{r} = r\mathbf{e}_r + z\mathbf{E}_z = x\mathbf{E}_x + y\mathbf{E}_y + z\mathbf{E}_z \, .$$

To calculate the (absolute) velocity vector $\mathbf{v}$ of the particle, we differentiate $\mathbf{r}(t)$:

$$\mathbf{v} = \frac{d\mathbf{r}}{dt} = \frac{dr}{dt}\mathbf{e}_r + r\frac{d\mathbf{e}_r}{dt} + \frac{dz}{dt}\mathbf{E}_z \, .$$

Now, using the chain rule, $\dot{\mathbf{e}}_r = \dot{\theta}\frac{d\mathbf{e}_r}{d\theta} = \dot{\theta}\mathbf{e}_\theta$. Also,

$$\frac{d\mathbf{e}_\theta}{d\theta} = -\mathbf{e}_r \, , \qquad \frac{d\mathbf{e}_r}{d\theta} = \mathbf{e}_\theta \, .$$

It follows that

$$\begin{aligned}
\mathbf{v} &= \frac{dr}{dt}\mathbf{e}_r + r\frac{d\theta}{dt}\mathbf{e}_\theta + \frac{dz}{dt}\mathbf{E}_z \\
&= \frac{dx}{dt}\mathbf{E}_x + \frac{dy}{dt}\mathbf{E}_y + \frac{dz}{dt}\mathbf{E}_z \, .
\end{aligned}$$

To calculate the (absolute) acceleration vector $\mathbf{a}$, we differentiate $\mathbf{v}$ with respect to time:

$$\mathbf{a} = \frac{d\mathbf{v}}{dt} = \frac{d}{dt}\left(\frac{dr}{dt}\mathbf{e}_r\right) + \frac{d}{dt}\left(r\frac{d\theta}{dt}\mathbf{e}_\theta\right) + \frac{d^2z}{dt^2}\mathbf{E}_z \, .$$

Using the chain rule to determine the time derivatives of the vectors $\mathbf{e}_r$ and $\mathbf{e}_\theta$, and after collecting terms in the expressions for $\mathbf{a}$, the final form of the results are obtained:

$$\begin{aligned}
\mathbf{a} &= \left(\frac{d^2r}{dt^2} - r\left(\frac{d\theta}{dt}\right)^2\right)\mathbf{e}_r + \left(r\frac{d^2\theta}{dt^2} + 2\frac{dr}{dt}\frac{d\theta}{dt}\right)\mathbf{e}_\theta + \frac{d^2z}{dt^2}\mathbf{E}_z \\
&= \frac{d^2x}{dt^2}\mathbf{E}_x + \frac{d^2y}{dt^2}\mathbf{E}_y + \frac{d^2z}{dt^2}\mathbf{E}_z \, .
\end{aligned}$$

We have also included the representations for the velocity and acceleration vectors in Cartesian coordinates to emphasize the fact that the values of these vectors do not depend on the coordinate system used.

2.2.1   Common Errors

In my experience, the most common error with using cylindrical polar coordinates is to write $\mathbf{r} = r\mathbf{e}_r + \theta\mathbf{e}_\theta + z\mathbf{E}_z$. *This is not true.*

Another mistake is to differentiate $\mathbf{e}_r$ and $\mathbf{e}_\theta$ incorrectly with respect to time.

Last, but not least, many people presume that all of the results presented here apply when θ is taken to be clockwise positive. Alas, this is not the case.

2.3   Kinetics of a Particle

Consider a particle of mass m. Let $\mathbf{F}$ denote the resultant external force acting on the particle, and let $\mathbf{G} = m\mathbf{v}$ be the linear momentum of the particle. Euler's first law (also known as Newton's second law, or the balance of linear momentum) postulates that

$$\mathbf{F} = \frac{d\mathbf{G}}{dt} = m\mathbf{a}.$$

With respect to a Cartesian basis $\mathbf{F} = m\mathbf{a}$ is equivalent to 3 scalar equations:

$$F_x = ma_x = m\ddot{x}, \quad F_y = ma_y = m\ddot{y}, \quad F_z = ma_z = m\ddot{z},$$

where $\mathbf{F} = F_x\mathbf{E}_x + F_y\mathbf{E}_y + F_z\mathbf{E}_z$ and $\mathbf{a} = a_x\mathbf{E}_x + a_y\mathbf{E}_y + a_z\mathbf{E}_z$.

With respect to a cylindrical polar coordinate system, the single vector equation $\mathbf{F} = m\mathbf{a}$ is equivalent to 3 scalar equations:

$$(\mathbf{F} = m\mathbf{a}) \cdot \mathbf{e}_r \quad : \quad F_r = m\left(\frac{d^2r}{dt^2} - r\left(\frac{d\theta}{dt}\right)^2\right),$$

$$(\mathbf{F} = m\mathbf{a}) \cdot \mathbf{e}_\theta \quad : \quad F_\theta = m\left(r\frac{d^2\theta}{dt^2} + 2\frac{dr}{dt}\frac{d\theta}{dt}\right),$$

$$(\mathbf{F} = m\mathbf{a}) \cdot \mathbf{E}_z \quad : \quad F_z = m\frac{d^2z}{dt^2}.$$

Finally, we recall for emphasis the relations

$$\mathbf{e}_r = \cos(\theta)\mathbf{E}_x + \sin(\theta)\mathbf{E}_y,$$

$$\mathbf{e}_\theta = -\sin(\theta)\mathbf{E}_x + \cos(\theta)\mathbf{E}_y,$$

$$\mathbf{E}_x = \cos(\theta)\mathbf{e}_r - \sin(\theta)\mathbf{e}_\theta,$$

$$\mathbf{E}_y = \sin(\theta)\mathbf{e}_r + \cos(\theta)\mathbf{e}_\theta.$$

You will use these relations several hundred times in an undergraduate engineering course.

2.4   The Planar Pendulum

The planar pendulum is a classical problem in mechanics. As shown in Figure 2.2, a particle of mass m is suspended from a fixed point O either by an inextensible massless string or rigid massless rod of length L. The particle is free to move on a plane ($z = 0$), and during its motion a vertical gravitational force $-mg\mathbf{E}_y$ acts on the particle.

One asks the questions, what are the equations governing the motion of the particle, and what is the tension in the string or rod?

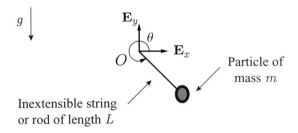

FIGURE 2.2. The planar pendulum

2.4.1 Kinematics

We begin by establishing some kinematical results. We note that $\mathbf{r} = L\mathbf{e}_r$. Differentiating with respect to t, and noting that L is constant, gives us the velocity $\mathbf{v}$. Similarly, we obtain $\mathbf{a}$ from $\mathbf{v}$:

$$\mathbf{v} = L\frac{d\mathbf{e}_r}{dt} = L\frac{d\theta}{dt}\mathbf{e}_\theta\,,$$

$$\mathbf{a} = L\frac{d^2\theta}{dt^2}\mathbf{e}_\theta + L\frac{d\theta}{dt}\frac{d\mathbf{e}_\theta}{dt} = L\frac{d^2\theta}{dt^2}\mathbf{e}_\theta - L\left(\frac{d\theta}{dt}\right)^2\mathbf{e}_r\,.$$

Alternatively, one can get these results by substituting $r = L$ and $z = 0$ in the general expressions recorded in Section 2. I don't recommend this approach, it emphasizes memorization.

2.4.2 Forces

Next, as shown in Figure 2.3, we draw a free-body diagram. There is a tension force $-T\mathbf{e}_r$ and a normal force $N\mathbf{E}_z$ acting on a particle. The role of the tension force is to ensure that the distance of the particle from the origin is L, while the normal force ensures that there is no motion in the direction of $\mathbf{E}_z$. These two forces are known as constraint forces. They are indeterminate (we need to use $\mathbf{F} = m\mathbf{a}$ to determine them). One should also note that the gravitational force has the representations

$$-mg\mathbf{E}_y = -mg\sin(\theta)\mathbf{e}_r - mg\cos(\theta)\mathbf{e}_\theta\,.$$

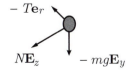

FIGURE 2.3. Free-body diagram for the planar pendulum

2.4.3 Balance Law

The third step is to write down the balance of linear momentum ($\mathbf{F} = m\mathbf{a}$):

$$-T\mathbf{e}_r + N\mathbf{E}_z - mg\mathbf{E}_y = mL\frac{d^2\theta}{dt^2}\mathbf{e}_\theta - mL\left(\frac{d\theta}{dt}\right)^2\mathbf{e}_r\,.$$

We obtain 3 scalar equations from this vector equation:

$$mL\frac{d^2\theta}{dt^2} = -mg\cos(\theta)\,,\quad T = mL\left(\frac{d\theta}{dt}\right)^2 - mg\sin(\theta)\,,\quad N = 0\,.$$

2.4.4 Analysis

The first of these equations is a second-order differential equation for $\theta(t)$:

$$mL\frac{d^2\theta}{dt^2} = -mg\cos(\theta)\,.$$

Given the initial conditions $\theta(t_0)$ and $\dot\theta(t_0)$, one can solve this equation and determine the motion of the particle. Next, the second equation gives the tension T in the string or rod, once $\theta(t)$ is known:

$$-T\mathbf{e}_r = -\left(mL\left(\frac{d\theta}{dt}\right)^2 - mg\sin(\theta)\right)\mathbf{e}_r\,.$$

For a string, it is normally assumed that $T > 0$, and for some motions of the string T will become negative. In this case, the particle behaves as if it were free to move on the plane and $r \neq L$. Lastly, the normal force $N\mathbf{E}_z$ is zero in this problem.

2.5 Summary

In this chapter, the cylindrical polar coordinate system r, θ, z was introduced. To assist with certain expressions, the vectors $\mathbf{e}_r = \cos(\theta)\mathbf{E}_x + \sin(\theta)\mathbf{E}_y$ and $\mathbf{e}_\theta = -\sin(\theta)\mathbf{E}_x + \cos(\theta)\mathbf{E}_y$ were introduced. It was also shown that the position vector of a particle has the representations

$$\begin{aligned}
\mathbf{r} &= r\mathbf{e}_r + z\mathbf{E}_z = \sqrt{x^2 + y^2}\mathbf{e}_r + z\mathbf{E}_z \\
&= r\cos(\theta)\mathbf{E}_x + r\sin(\theta)\mathbf{E}_y + z\mathbf{E}_z \\
&= x\mathbf{E}_x + y\mathbf{E}_y + z\mathbf{E}_z\,.
\end{aligned}$$

By differentiating $\mathbf{r}$ with respect to time, the velocity and acceleration vectors were obtained. These vectors have the representations

$$\mathbf{v} = \frac{d\mathbf{r}}{dt} = \dot{r}\mathbf{e}_r + r\dot\theta\mathbf{e}_\theta + \dot{z}\mathbf{E}_z$$

$$= \dot{x}\mathbf{E}_x + \dot{y}\mathbf{E}_y + \dot{z}\mathbf{E}_z \,,$$

$$\mathbf{a} = \frac{d\mathbf{v}}{dt} = \left(\ddot{r} - r\dot{\theta}^2\right)\mathbf{e}_r + \left(r\ddot{\theta} + 2\dot{r}\dot{\theta}\right)\mathbf{e}_\theta + \ddot{z}\mathbf{E}_z$$

$$= \ddot{x}\mathbf{E}_x + \ddot{y}\mathbf{E}_y + \ddot{z}\mathbf{E}_z \,.$$

To establish these results, the chain rule and the important identities $\dot{\mathbf{e}}_r = \dot{\theta}\mathbf{e}_\theta$ and $\dot{\mathbf{e}}_\theta = -\dot{\theta}\mathbf{e}_r$ were used.

Using a cylindrical polar coordinate system, $\mathbf{F} = m\mathbf{a}$ can be written as three scalar equations:

$$F_r = m\left(\ddot{r} - r\dot{\theta}^2\right),$$

$$F_\theta = m\left(r\ddot{\theta} + 2\dot{r}\dot{\theta}\right),$$

$$F_z = m\ddot{z}.$$

These equations were illustrated using the example of the planar pendulum.

2.6   Exercises

The following short exercises are intended to assist you in reviewing Chapter 2.

2.1 Using Figure 2.1, verify that $\mathbf{e}_r = \cos(\theta)\mathbf{E}_x + \sin(\theta)\mathbf{E}_y$ and $\mathbf{e}_\theta = -\sin(\theta)\mathbf{E}_x + \cos(\theta)\mathbf{E}_y$. Then, by considering cases where $\mathbf{e}_r$ lies in the second, third, and fourth quadrants, verify that these definitions are valid for all values of θ.

2.2 Starting from the definitions $\mathbf{e}_r = \cos(\theta)\mathbf{E}_x + \sin(\theta)\mathbf{E}_y$ and $\mathbf{e}_\theta = -\sin(\theta)\mathbf{E}_x + \cos(\theta)\mathbf{E}_y$, show that $\dot{\mathbf{e}}_r = \dot{\theta}\mathbf{e}_\theta$ and $\dot{\mathbf{e}}_\theta = -\dot{\theta}\mathbf{e}_r$. In addition, verify that $\mathbf{E}_x = \cos(\theta)\mathbf{e}_r - \sin(\theta)\mathbf{e}_\theta$ and $\mathbf{E}_y = \sin(\theta)\mathbf{e}_r + \cos(\theta)\mathbf{e}_\theta$.

2.3 Calculate the velocity vectors of particles whose position vectors are $10\mathbf{e}_r$ and $5\mathbf{e}_r + t\mathbf{E}_z$, where $\theta = \pi t$. Why do all of these particles move with constant speed $\|\mathbf{v}\|$, yet have a non–zero acceleration?

2.4 The position vector of a particle of mass m which is placed at the end of a rotating, telescoping rod is $\mathbf{r} = 6t\mathbf{e}_r$ where $\theta = 10t + 5$ (radians). Calculate the velocity and acceleration vectors of the particle, and determine the force $\mathbf{F}$ needed to sustain the motion of particle. What is the force that the particle exerts on the telescoping rod?

2.5 In solving a problem, one person uses cylindrical polar coordinates while another uses Cartesian coordinates. To check that their answers are identical, they need to examine the relationship between the Cartesian and cylindrical polar components of a certain vector, say $\mathbf{b}$. To this end, show that

$$b_x = \mathbf{b} \cdot \mathbf{E}_x = b_r \cos(\theta) - b_\theta \sin(\theta)\,, \quad b_y = \mathbf{b} \cdot \mathbf{E}_y = b_r \sin(\theta) + b_\theta \cos(\theta)\,.$$

2.6 Consider the projectile problem discussed in Section 5 of Chapter 1. Using a cylindrical polar coordinate system, show that the equations governing the motion of the particle are

$$m\ddot{r} - mr\dot{\theta}^2 = -mg\sin(\theta)\,, \quad mr\ddot{\theta} + 2m\dot{r}\dot{\theta} = -mg\cos(\theta)\,, \quad m\ddot{z} = 0\,.$$

Notice that, in contrast to using Cartesian coordinates to determine the governing equations, solving these differential equations is non-trivial.

2.7 Consider a spherical bead of mass m and radius R that is placed inside a long cylindrical tube. The inner radius of the tube is R, and the tube is pivoted so that it rotates in a horizontal plane. Furthermore, the contact between the tube and the bead is smooth. Here, the bead is modeled as a particle of mass m. Now suppose that the tube is whirled at a constant angular speed Ω (radians per second). The whirling motion of the tube is such that the velocity vector of the bead is $\mathbf{v} = \dot{r}\mathbf{e}_r + \Omega r \mathbf{e}_\theta$. Show that the equation governing the motion of the bead is $\ddot{r} - \Omega^2 r = 0$ and the force exerted by the tube on the particle is $mg\mathbf{E}_z + 2m\dot{r}\Omega\mathbf{e}_\theta$.

2.8 Consider the case where the bead is initially at rest relative to the whirling tube at a location $r_0 = L$. Using the solution to the differential equation $\ddot{r} - \Omega^2 r = 0$ recorded in Section 5.4 of Appendix A, show that, unless $L = 0$, the bead discussed in the previous exercise will eventually exit the whirling tube.

3
Particles and Space Curves

TOPICS

In this chapter we discuss the differential geometry of space curves (a curve embedded in Euclidean three-space $\mathcal{E}^3$). In particular, we introduce the Serret-Frenet basis vectors $\{\mathbf{e}_t, \mathbf{e}_n, \mathbf{e}_b\}$. This is followed by the derivation of an elegant set of relations describing the rate of change of the tangent $\mathbf{e}_t$, principal normal $\mathbf{e}_n$, and binormal $\mathbf{e}_b$ vectors. Several examples of space curves are then discussed. We end the chapter with some applications to the mechanics of particles. Subsequent chapters will also discuss several examples.

3.1 The Serret-Frenet Triad

For many problems, such as the motion of a particle on a straight line, or on a plane, it suffices to know how to manipulate Cartesian or polar coordinates. However, these systems become cumbersome for many problems. In general, the path of the particle is a curve in space. Often this curve is prescribed, as in the case of a particle moving on a circular path, or else it may lie on a surface. More often than not, this curve is not known a priori.

We now turn to defining the Serret-Frenet basis vectors $\{\mathbf{e}_t, \mathbf{e}_n, \mathbf{e}_b\}$ for a point P of this space curve.[1]

Consider a fixed curve[2] $\mathcal{C}$ that is embedded in $\mathcal{E}^3$. The curve $\mathcal{C}$ is often known as a *space curve*. Let the position vector of a point $P \in \mathcal{C}$ be denoted by $\mathbf{r}$. This vector has the representation

$$\mathbf{r} = x\mathbf{E}_x + y\mathbf{E}_y + z\mathbf{E}_z,$$

where x, y, and z are the usual Cartesian coordinates and $\mathbf{E}_x$, $\mathbf{E}_y$, and $\mathbf{E}_z$ are the orthonormal basis vectors associated with these coordinates.

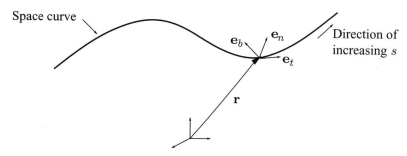

FIGURE 3.1. A space curve and the Serret-Frenet triad at one of its points

Associated with the curve $\mathcal{C}$, we can define an arc-length parameter s, where by definition,

$$\left(\frac{ds}{dt}\right)^2 = \frac{d\mathbf{r}}{dt} \cdot \frac{d\mathbf{r}}{dt} = \frac{dx}{dt}\frac{dx}{dt} + \frac{dy}{dt}\frac{dy}{dt} + \frac{dz}{dt}\frac{dz}{dt}.$$

This parameter uniquely identifies a point P of $\mathcal{C}$, and we can use it to obtain a different representation of the position vector of a point on the curve:

$$\mathbf{r} = \hat{\mathbf{r}}(s).$$

Consider two points P and P' of $\mathcal{C}$, where the position vectors of P and P' are $\hat{\mathbf{r}}(s)$ and $\hat{\mathbf{r}}(s + \Delta s)$, respectively. We define the vector $\mathbf{e}_t$ as

$$\mathbf{e}_t = \hat{\mathbf{e}}_t(s) = \frac{d\mathbf{r}}{ds} = \lim_{\Delta s \to 0} \frac{\hat{\mathbf{r}}(s + \Delta s) - \hat{\mathbf{r}}(s)}{\Delta s}.$$

---

[1] These triads and formulae for their rates of change were established by Jean-Frédéric Frenet (1816–1900) in 1847 (see [26]) and Joseph Alfred Serret (1819–1885) in 1851 (see [55]).

[2] One can extend all of our forthcoming discussion to curves that are moving in space, for example, a curve in the form of a circle whose radius changes with time. In this case $\mathbf{r} = \mathbf{r}(s, t)$, and the Serret-Frenet triad is obtained by the partial differentiation of $\mathbf{r}$ with respect to s. Such an extension is well-known and needed for discussing the theory of rods, but is beyond the scope of an undergraduate engineering dynamics course.

Since $\hat{\mathbf{r}}(s + \Delta s) - \hat{\mathbf{r}}(s) \to \Delta s \mathbf{e}_t$ as $\Delta s \to 0,^3$ $\mathbf{e}_t$ is known as the unit tangent vector to $\mathcal{C}$ at the point P (see Figure 3.1).

Consider the second derivative of $\mathbf{r}$ with respect to s:

$$\frac{d^2\mathbf{r}}{ds^2} = \frac{d\mathbf{e}_t}{ds}.$$

After noting that $\mathbf{e}_t \cdot \mathbf{e}_t = 1$, and then differentiating this relation, we find that $\frac{d\mathbf{e}_t}{ds}$ is perpendicular to $\mathbf{e}_t$. We define κ to be the magnitude of $\frac{d\mathbf{e}_t}{ds}$ and $\mathbf{e}_n$ to be its direction:

$$\kappa \mathbf{e}_n = \frac{d\mathbf{e}_t}{ds}.$$

The vector $\mathbf{e}_n$ is known as the (unit) principal normal vector to $\mathcal{C}$ at the point P, and the scalar κ is known as the curvature of $\mathcal{C}$ at the point $P.^4$ It should be noted that $\mathbf{e}_n$ and κ are functions of s:

$$\kappa = \hat{\kappa}(s), \quad \mathbf{e}_n = \hat{\mathbf{e}}_n(s).$$

Often an additional variable ρ, which is known as the radius of curvature, is defined:

$$\rho = \hat{\rho}(s) = \frac{1}{\kappa}.$$

For the degenerate case where $\frac{d\mathbf{e}_t}{ds} = \mathbf{0}$ for a particular s, the curvature κ is 0, and the vector $\mathbf{e}_n$ is not uniquely defined. In this case, one usually defines $\mathbf{e}_n$ to be a unit vector perpendicular to $\mathbf{e}_t$. The most common case of this occurrence is when the curve $\mathcal{C}$ is a straight line.

The final vector of interest is $\mathbf{e}_b$, and it is defined by

$$\mathbf{e}_b = \hat{\mathbf{e}}_b(s) = \mathbf{e}_t \times \mathbf{e}_n.$$

Clearly, $\mathbf{e}_b$ is a unit vector. It is known as the (unit) binormal vector to $\mathcal{C}$ at the point P.

It should be noted that the set of vectors $\{\mathbf{e}_t, \mathbf{e}_n, \mathbf{e}_b\}$ are defined for each point of $\mathcal{C}$, are orthonormal, and, because $\mathbf{e}_b \cdot (\mathbf{e}_t \times \mathbf{e}_n) = 1$, form a right-handed set. We shall refer to the set $\{\mathbf{e}_t, \mathbf{e}_n, \mathbf{e}_b\}$ as the Serret-Frenet triad. Because this set is orthonormal it may be used as a basis for $\mathcal{E}^3$. That is, given any vector $\mathbf{b}$, one has the following representations:

$$\mathbf{b} = b_t \mathbf{e}_t + b_n \mathbf{e}_n + b_b \mathbf{e}_b = b_x \mathbf{E}_x + b_y \mathbf{E}_y + b_z \mathbf{E}_z,$$

where $b_t = \mathbf{b} \cdot \mathbf{e}_t$, etc.

---

[3] This may be easily seen by sketching the curve and the position vectors of P and P', and then taking the limit as Δs tends to 0.

[4] Often the convention that κ is nonnegative is adopted. This allows the convenient identification that $\mathbf{e}_n$ points in the direction of $\frac{d\mathbf{e}_t}{ds}$. We shall adhere strictly to this convention.

At a particular s, the plane defined by the vectors $\mathbf{e}_t$ and $\mathbf{e}_n$ is known as the *osculating plane*, and the plane defined by the vectors $\mathbf{e}_t$ and $\mathbf{e}_b$ is known as the *rectifying plane*. These planes will, in general, depend on the particular point P of $\mathcal{C}$.[5]

3.2   The Serret-Frenet Formulae

These three formulae relate the rate of change of the vectors $\mathbf{e}_t$, $\mathbf{e}_n$, and $\mathbf{e}_b$ with respect to the arc-length parameter s to the set of vectors $\{\mathbf{e}_t, \mathbf{e}_n, \mathbf{e}_b\}$.

Consider the vector $\mathbf{e}_t$. Recalling one of the previous results, the first of the desired formulae is recorded:

$$\frac{d\mathbf{e}_t}{ds} = \kappa \mathbf{e}_n \, .$$

It is convenient to consider next the vector $\mathbf{e}_b$. Since this vector is a unit vector, it cannot change along its own length. Mathematically, we see this as follows:

$$\mathbf{e}_b = \mathbf{e}_t \times \mathbf{e}_n \quad \Longrightarrow \quad \mathbf{e}_b \cdot \mathbf{e}_b = 1$$
$$\Longrightarrow \quad \frac{d\mathbf{e}_b}{ds} \cdot \mathbf{e}_b = 0 \, .$$

Consequently, $\frac{d\mathbf{e}_b}{ds}$ has components only in the directions of $\mathbf{e}_t$ and $\mathbf{e}_n$. Let's first examine what the component of $\frac{d\mathbf{e}_b}{ds}$ is in the $\mathbf{e}_t$ direction:

$$\mathbf{e}_t \cdot \mathbf{e}_b = 0 \quad \Longrightarrow \quad \frac{d\mathbf{e}_t}{ds} \cdot \mathbf{e}_b + \frac{d\mathbf{e}_b}{ds} \cdot \mathbf{e}_t = 0$$
$$\Longrightarrow \quad \kappa \mathbf{e}_n \cdot \mathbf{e}_b + \frac{d\mathbf{e}_b}{ds} \cdot \mathbf{e}_t = 0$$
$$\Longrightarrow \quad \frac{d\mathbf{e}_b}{ds} \cdot \mathbf{e}_t = 0 \, .$$

It now follows that $\frac{d\mathbf{e}_b}{ds}$ is parallel to $\mathbf{e}_n$. Consequently, we define

$$\frac{d\mathbf{e}_b}{ds} = -\tau \mathbf{e}_n \, ,$$

where $\tau = \hat{\tau}(s)$ is the torsion of the curve $\mathcal{C}$ at the particular point P corresponding to the value of s. The negative sign in the above formula is conventional.

---

[5]It is beyond our purposes to present an additional discussion on these planes. The interested reader is referred to an introductory text on differential geometry, several of which are available. We mention in particular Kreyszig [35], Spivak [59], and Struik [63].

To obtain the final Serret-Frenet formula for $\frac{d\mathbf{e}_n}{ds}$, we do a direct calculation:

$$
\begin{aligned}
\frac{d\mathbf{e}_n}{ds} &= \frac{d}{ds}(\mathbf{e}_b \times \mathbf{e}_t) = \frac{d\mathbf{e}_b}{ds} \times \mathbf{e}_t + \mathbf{e}_b \times \frac{d\mathbf{e}_t}{ds} \\
&= (-\tau\mathbf{e}_n) \times \mathbf{e}_t + \mathbf{e}_b \times (\kappa\mathbf{e}_n).
\end{aligned}
$$

After simplifying this result by evaluating the cross products, we obtain

$$
\frac{d\mathbf{e}_n}{ds} = -\kappa\mathbf{e}_t + \tau\mathbf{e}_b.
$$

The Serret-Frenet formulae can be conveniently summarized as

$$
\begin{bmatrix} \frac{d\mathbf{e}_t}{ds} \\ \frac{d\mathbf{e}_n}{ds} \\ \frac{d\mathbf{e}_b}{ds} \end{bmatrix} = \begin{bmatrix} 0 & \kappa & 0 \\ -\kappa & 0 & \tau \\ 0 & -\tau & 0 \end{bmatrix} \begin{bmatrix} \mathbf{e}_t \\ \mathbf{e}_n \\ \mathbf{e}_b \end{bmatrix}.
$$

One can define what is often referred to as the Darboux[6] vector $\boldsymbol{\omega}_{\mathrm{SF}}$,

$$
\boldsymbol{\omega}_{\mathrm{SF}} = \kappa\mathbf{e}_b + \tau\mathbf{e}_t.
$$

Using the Darboux vector, the Serret-Frenet relations can also be written in the form

$$
\frac{d\mathbf{e}_i}{ds} = \boldsymbol{\omega}_{\mathrm{SF}} \times \mathbf{e}_i,
$$

where $i = t$, n, or b.

3.3 Examples of Space Curves

We proceed to discuss four examples of spaces curves: a plane curve, a circle, a space curve parametrized by x, and a circular helix. The degenerate case of a straight line is discussed in conjunction with the plane curve.

3.3.1 A Curve on a Plane

As shown in Figure 3.2,[7] consider a curve on a plane in $\mathcal{E}^3$. The plane is defined by the relation $z = z_0$, and the curve is defined by the intersection of two 2-dimensional surfaces:

$$
z = z_0, \quad y = f(x),
$$

where we shall assume that f is as smooth as necessary. A specific example will be presented in Section 5.

---

[6]Gaston Darboux (1842–1917) was a French mathematician who wrote an authoritative four-volume treatise on differential geometry, which was published between 1887 and 1896 (see [19]).

[7]The convention that $\kappa \geq 0$ has been used to define the vector $\mathbf{e}_n$.

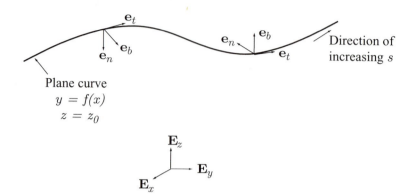

FIGURE 3.2. An example of a plane curve

The position vector of a point P on this curve is

$$\mathbf{r} = x\mathbf{E}_x + y\mathbf{E}_y + z\mathbf{E}_z = x\mathbf{E}_x + f(x)\mathbf{E}_y + z_0\mathbf{E}_z \ .$$

To determine the arc-length parameter s of the curve, we first note that

$$\frac{d\mathbf{r}}{dt} = \frac{dx}{dt}\mathbf{E}_x + \frac{df}{dx}\frac{dx}{dt}\mathbf{E}_y \ .$$

Consequently,

$$\left(\frac{ds}{dt}\right)^2 = \left(1 + \left(\frac{df}{dx}\right)^2\right)\left(\frac{dx}{dt}\right)^2 \ ,$$

or, assuming that s increases in the direction of increasing x,

$$\frac{ds}{dt} = \sqrt{\left(1 + \left(\frac{df}{dx}\right)^2\right)}\frac{dx}{dt} \ .$$

Integrating both sides of the above equation, we obtain

$$s = s(x) = \int_{x_0}^{x}\sqrt{1 + \left(\frac{df}{dx}\right)^2}\,du + s(x_0) \ .$$

As follows from our previous development, we should invert $s(x)$ to determine $x(s)$. However, since we prefer to keep the function $f(x)$ arbitrary, we shall express the results as functions of x.

To determine the tangent vector, we recall its definition and use the chain rule:

$$\mathbf{e}_t = \mathbf{e}_t(x) = \frac{d\mathbf{r}}{ds} = \frac{d\mathbf{r}}{dx}\frac{dx}{ds} = \frac{1}{\sqrt{1 + \left(\frac{df}{dx}\right)^2}}\left(\mathbf{E}_x + \frac{df}{dx}\mathbf{E}_y\right) \ ,$$

where we have also used the identity $\frac{dx}{ds} = \left(\frac{ds}{dx}\right)^{-1}$. The expected unit magnitude of $\mathbf{e}_t$ should be noted.

The principal normal vector $\mathbf{e}_n$ and the curvature κ are determined by evaluating the derivative of $\mathbf{e}_t$ with respect to s:

$$\kappa \mathbf{e}_n = \frac{d\mathbf{e}_t}{ds} = \frac{d\mathbf{e}_t}{dx}\frac{dx}{ds} = \frac{d\mathbf{e}_t}{dx}\left(\frac{ds}{dx}\right)^{-1} .$$

Omitting the details of the calculation, after some subsequent rearrangement we obtain

$$\kappa \mathbf{e}_n = \frac{\frac{d^2 f}{dx^2}}{\left(1 + \left(\frac{df}{dx}\right)^2\right)^2}\left(\mathbf{E}_y - \frac{df}{dx}\mathbf{E}_x\right) .$$

Recalling that $\mathbf{e}_n$ is a unit vector and that κ is positive, the final results are obtained:

$$\kappa \;=\; \kappa(x) = \frac{\left|\frac{d^2 f}{dx^2}\right|}{\left(\sqrt{1 + \left(\frac{df}{dx}\right)^2}\right)^3} ,$$

$$\mathbf{e}_n \;=\; \mathbf{e}_n(x) = \frac{\operatorname{sgn}\left(\frac{d^2 f}{dx^2}\right)}{\sqrt{1 + \left(\frac{df}{dx}\right)^2}}\left(\mathbf{E}_y - \frac{df}{dx}\mathbf{E}_x\right) .$$

Here, $\operatorname{sgn}(a) = 1$ if $a > 0$ and -1 if $a < 0$.

Finally, the binormal vector $\mathbf{e}_b$ may be determined:

$$\mathbf{e}_b = \mathbf{e}_t \times \mathbf{e}_n = \operatorname{sgn}\left(\frac{d^2 f}{dx^2}\right)\mathbf{E}_z ,$$

and, since this vector is a piecewise constant, the torsion of the curve is

$$\tau = 0 .$$

Returning briefly to Figure 3.2, you may have noticed that for certain segments of the curve $\mathbf{e}_b = \mathbf{E}_z$, while for others $\mathbf{e}_b = -\mathbf{E}_z$. The points where this transition occurs are those where $\frac{d^2 f}{dx^2} = 0$. At these points, $\kappa = 0$, and $\mathbf{e}_n$ is not defined by the Serret-Frenet formula $\frac{d\mathbf{e}_t}{ds} = \kappa \mathbf{e}_n$.

For the plane curve, many texts use a particular representation of the tangent and normal vectors by defining an angle $\beta = \beta(s)$ as follows:

$$\mathbf{e}_t = \cos\left(\beta(s)\right)\mathbf{E}_x + \sin\left(\beta(s)\right)\mathbf{E}_y ,$$

$$\mathbf{e}_n = \cos\left(\beta(s)\right)\mathbf{E}_y - \sin\left(\beta(s)\right)\mathbf{E}_x .$$

Notice that $\mathbf{e}_t$ and $\mathbf{e}_n$ are unit vectors, as expected. By differentiating these expressions with respect to s, one finds that

$$\kappa = \frac{d\beta}{ds} \,.$$

Consequently, κ can be interpreted as a rate of rotation of the vectors $\mathbf{e}_t$ and $\mathbf{e}_n$ about $\mathbf{e}_b = \pm\mathbf{E}_z$.

The Straight Line

A special case of the plane curve arises when $f(x) = ax + b$, where a and b are constants. In this case, we find from above that

$$\mathbf{e}_t = \mathbf{e}_t(x) = \frac{1}{\sqrt{1 + \left(\frac{df}{dx}\right)^2}} \left(\mathbf{E}_x + \frac{df}{dx}\mathbf{E}_y\right) = \frac{1}{\sqrt{1 + a^2}} \left(\mathbf{E}_x + a\mathbf{E}_y\right) .$$

It should be clear that we are assuming that $\frac{ds}{dx} = \sqrt{1 + \left(\frac{df}{dx}\right)^2}$, as opposed to $-\sqrt{1 + \left(\frac{df}{dx}\right)^2}$. Turning to the principal normal vector, because $\frac{d\mathbf{e}_t}{ds} = \mathbf{0}$, the curvature $\kappa = 0$ and $\mathbf{e}_n$ is not defined. For consistency, it is convenient to choose $\mathbf{e}_n$ to be perpendicular to $\mathbf{e}_t$. The binormal vector is then defined by $\mathbf{e}_b = \mathbf{e}_t \times \mathbf{e}_n$.

3.3.2 A Space Curve Parametrized by x

Consider a curve $\mathcal{C}$ in $\mathcal{E}^3$. Suppose that the curve is defined by the intersection of the two 2-dimensional surfaces

$$z = g(x), \quad y = f(x),$$

where we shall assume that f and g are as smooth as necessary. The plane curve discussed in Section 3.1 can be considered a special case of this one. The position vector of a point P on this curve is

$$\mathbf{r} = x\mathbf{E}_x + f(x)\mathbf{E}_y + g(x)\mathbf{E}_z .$$

The arc-length parameter s may be determined in a manner similar to the previous developments:

$$s = s(x) = \int_{x_0}^{x} \sqrt{1 + \left(\frac{df}{dx}\right)^2 + \left(\frac{dg}{dx}\right)^2} \, du - s(x_0) .$$

As in Section 3.1, we shall express all of our results as functions of x and assume that the function $x(s)$ is available to us. It follows that we can later express all of our results as functions of s if required.

To calculate the tangent vector, we use the chain rule as before:

$$
\begin{aligned}
\mathbf{e}_t &= \mathbf{e}_t(x) = \frac{d\mathbf{r}}{ds} = \frac{d\mathbf{r}}{dx}\frac{dx}{ds} \\
&= \frac{1}{\sqrt{1 + \left(\frac{df}{dx}\right)^2 + \left(\frac{dg}{dx}\right)^2}}\left(\mathbf{E}_x + \frac{df}{dx}\mathbf{E}_y + \frac{dg}{dx}\mathbf{E}_z\right).
\end{aligned}
$$

The principal normal vector $\mathbf{e}_n$ and the curvature κ are determined by evaluating the derivative of $\mathbf{e}_t$ with respect to s:

$$
\kappa\mathbf{e}_n = \frac{d\mathbf{e}_t}{ds} = \frac{d\mathbf{e}_t}{dx}\frac{dx}{ds} = \frac{d\mathbf{e}_t}{dx}\left(\frac{ds}{dx}\right)^{-1}.
$$

Finally, the binormal vector $\mathbf{e}_b$ may be determined using its definition. We omit details of the expressions for the principal normal and binormal vectors, curvature κ, and the torsion τ. They may be obtained using the quoted relations, and their specific general forms are not of further interest here.

3.3.3 A Circle on a Plane

As shown in Figure 3.3, consider a curve in the form of a circle that lies on a plane in $\mathcal{E}^3$. The plane is defined by the relation $z = z_0$, and the curve is defined by the intersection of two 2-dimensional surfaces:

$$
z = z_0, \quad r = R = \sqrt{x^2 + y^2}.
$$

In addition, it is convenient to recall the relations

$$
\theta = \tan^{-1}\left(\frac{y}{x}\right),
$$

$$
\mathbf{e}_r = \cos(\theta)\mathbf{E}_x + \sin(\theta)\mathbf{E}_y, \quad \mathbf{e}_\theta = \cos(\theta)\mathbf{E}_y - \sin(\theta)\mathbf{E}_x.
$$

The position vector of a point P on this curve is

$$
\mathbf{r} = x\mathbf{E}_x + y\mathbf{E}_y + z_0\mathbf{E}_z = R\mathbf{e}_r + z_0\mathbf{E}_z.
$$

To determine the arc-length parameter s of the curve, we first note that

$$
\mathbf{v} = \frac{d\mathbf{r}}{dt} = R\frac{d\mathbf{e}_r}{dt} = R\frac{d\theta}{dt}\mathbf{e}_\theta.
$$

Consequently,

$$
\frac{ds}{dt} = R\frac{d\theta}{dt}.
$$

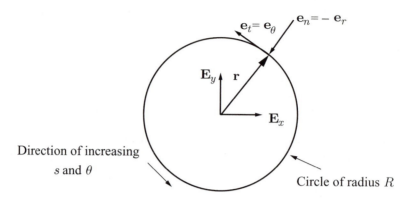

FIGURE 3.3. A circle of radius R

Here, we have assumed that s increases in the direction of increasing θ.[8] Integrating both sides of this equation, we obtain

$$s(\theta) = R(\theta - \theta_0) + s(\theta_0)\,.$$

Fortunately, we can invert the function $s(\theta)$ to solve for $\theta(s)$:

$$\theta(s) = \frac{1}{R}(s - s_0) + \theta(s_0)\,.$$

The previous result enables us to write

$$\mathbf{r} = \hat{\mathbf{r}}(s) = R\hat{\mathbf{e}}_r(s) + z_0\mathbf{E}_z\,,$$

where

$$\hat{\mathbf{e}}_r(s) = \cos(\theta(s))\mathbf{E}_x + \sin(\theta(s))\mathbf{E}_y\,.$$

It should be noted that the function $\hat{\mathbf{e}}_\theta(s)$ can be defined in a similar manner.

To determine the tangent vector, we differentiate $\mathbf{r}$ as a function of s:

$$\mathbf{e}_t = \hat{\mathbf{e}}_t(s) = \frac{d\mathbf{r}}{ds} = R\frac{d\mathbf{e}_r}{d\theta}\frac{d\theta}{ds} = R\mathbf{e}_\theta\frac{1}{R} = \mathbf{e}_\theta\,.$$

The expected unit magnitude of $\mathbf{e}_t$ should again be noted. The principal normal vector $\mathbf{e}_n$ and the curvature κ are determined by evaluating the derivative of $\mathbf{e}_t$ with respect to s:

$$\kappa\mathbf{e}_n = \hat{\kappa}(s)\hat{\mathbf{e}}_n(s) = \frac{d\mathbf{e}_t}{ds} = \frac{d\mathbf{e}_\theta}{d\theta}\frac{d\theta}{ds} = -\frac{1}{R}\mathbf{e}_r\,.$$

Recalling that $\mathbf{e}_n$ has unit magnitude and adopting the convention that $\kappa \geq 0$, we obtain

$$\kappa = \frac{1}{R}\,, \qquad \mathbf{e}_n = -\mathbf{e}_r\,.$$

---

[8]Otherwise, $\frac{ds}{d\theta} = -R$, and the forthcoming results need some minor modifications.

Finally, the binormal vector $\mathbf{e}_b$ may be determined:

$$\mathbf{e}_b = \mathbf{e}_t \times \mathbf{e}_n = \mathbf{E}_z \,.$$

Since this vector is constant, the torsion of the circular curve is, trivially,

$$\tau = 0 \,.$$

In summary, one has for a circle in the $\mathbf{E}_z$ plane,

$$\mathbf{e}_t = \mathbf{e}_\theta \,, \quad \mathbf{e}_n = -\mathbf{e}_r \,, \quad \mathbf{e}_b = \mathbf{E}_z \,, \quad \rho = R \,, \quad \tau = 0 \,.$$

Based on previous work with cylindrical polar coordinates, the tangent and normal vectors for this curve should have been anticipated.

3.3.4 A Circular Helix

Consider a curve in the form of a helix that is embedded in $\mathcal{E}^3$.[9] The helix is defined by the intersection of two 2-dimensional surfaces. One of these surfaces is defined by the relation

$$r = R \,,$$

where (r, θ, z) are the usual cylindrical polar coordinates:

$$r = \sqrt{x^2 + y^2} \,, \quad \theta = \tan^{-1}\left(\frac{y}{x}\right) \,.$$

Clearly, the surface $r = R$ is a cylinder. The second surface is defined by the equation

$$z = g(r, \theta) = \alpha r \theta \,.$$

This surface is known as a right helicoid.[10] Taking the intersection of the cylinder and helicoid one obtains a circular helix. An example of a circular helix is shown in Figure 3.4.

In Cartesian coordinates, the circular helix may be represented by

$$x = R \cos(\theta) \,, \quad y = R \sin(\theta) \,, \quad z = \alpha R \theta \,.$$

The position vector of a point P on the helix is

$$\mathbf{r} = x\mathbf{E}_x + y\mathbf{E}_y + z\mathbf{E}_z = R\mathbf{e}_r + \alpha R \theta \mathbf{E}_z \,,$$

where for convenience we have defined, as always,

$$\mathbf{e}_r = \cos(\theta)\mathbf{E}_x + \sin(\theta)\mathbf{E}_y \,, \quad \mathbf{e}_\theta = -\sin(\theta)\mathbf{E}_x + \cos(\theta)\mathbf{E}_y \,.$$

---

[9]This is an advanced example. According to Kreyszig [35], the circular helix is the only nontrivial example of a curve with constant torsion and constant curvature.

[10]See, for example, Section 2.2 and Fig. 3-4 of Struik [63].

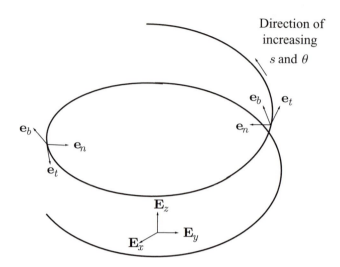

FIGURE 3.4. Examples of Serret-Frenet triads for a portion of a circular helix

To determine the arc-length parameter s of the curve we first note that

$$\frac{d\mathbf{r}}{dt} = R\frac{d\mathbf{e}_r}{dt} + \frac{d}{dt}\left(\alpha R\theta\mathbf{E}_z\right) = R\frac{d\theta}{dt}\mathbf{e}_\theta + \alpha R\frac{d\theta}{dt}\mathbf{E}_z .$$

Consequently,

$$\frac{ds}{dt} = R\sqrt{1+\alpha^2}\,\frac{d\theta}{dt} .$$

Here, we have assumed that s increases in the direction of increasing θ.[11] Integrating both sides of this equation we obtain

$$s(\theta) = R\sqrt{1+\alpha^2}\,(\theta - \theta_0) + s(\theta_0) .$$

Fortunately, as in the case of the plane circle, we can invert the function $s(\theta)$ to solve for $\theta(s)$:

$$\theta(s) = \frac{1}{R\sqrt{1+\alpha^2}}(s - s_0) + \theta(s_0) .$$

The previous result enables us to write $\mathbf{r}$ as a function of s:

$$\mathbf{r} = \hat{\mathbf{r}}(s) = R\hat{\mathbf{e}}_r(s) + \left(\frac{1}{R\sqrt{1+\alpha^2}}(s - s_0) + \theta(s_0)\right)R\alpha\mathbf{E}_z ,$$

where

$$\hat{\mathbf{e}}_r(s) = \cos\left(\theta(s)\right)\mathbf{E}_x + \sin\left(\theta(s)\right)\mathbf{E}_y .$$

---

[11]Otherwise, $\frac{ds}{d\theta} = -R\sqrt{1+\alpha^2}$, and the forthcoming results need some minor modifications.

It should be noted that the function $\hat{\mathbf{e}}_\theta(s)$ can be defined in a similar manner.

To determine the tangent vector, we differentiate $\mathbf{r}$ with respect to s:

$$\mathbf{e}_t = \hat{\mathbf{e}}_t(s) = \frac{d\mathbf{r}}{ds} = R\frac{d\mathbf{e}_r}{d\theta}\frac{d\theta}{ds} + R\alpha\frac{d\theta}{ds}\mathbf{E}_z = \frac{1}{\sqrt{1+\alpha^2}}\left(\mathbf{e}_\theta + \alpha\mathbf{E}_z\right).$$

The expected unit magnitude of $\mathbf{e}_t$ should again be noted. The principal normal vector $\mathbf{e}_n$ and the curvature κ are determined as usual by evaluating the derivative of $\mathbf{e}_t$ with respect to s:

$$\begin{aligned}
\kappa\mathbf{e}_n &= \hat{\kappa}(s)\hat{\mathbf{e}}_n(s) = \frac{d\mathbf{e}_t}{ds} = \frac{1}{\sqrt{1+\alpha^2}}\frac{d\mathbf{e}_\theta}{ds} + \frac{d}{ds}\left(\frac{\alpha}{\sqrt{1+\alpha^2}}\mathbf{E}_z\right) \\
&= -\frac{1}{R(1+\alpha^2)}\mathbf{e}_r.
\end{aligned}$$

Recalling that the vector $\mathbf{e}_n$ has unit magnitude and adopting the convention that $\kappa \geq 0$, we obtain

$$\kappa = \frac{1}{R(1+\alpha^2)}, \qquad \mathbf{e}_n = -\mathbf{e}_r.$$

Finally, the binormal vector $\mathbf{e}_b$ may be determined:

$$\mathbf{e}_b = \mathbf{e}_t \times \mathbf{e}_n = \frac{1}{\sqrt{1+\alpha^2}}\left(-\alpha\mathbf{e}_\theta + \mathbf{E}_z\right).$$

The binormal vector is not constant, and the (constant) torsion of the circular helix is

$$\tau = \frac{\alpha}{R(1+\alpha^2)}.$$

3.4 Application to Particle Mechanics

In applications of the Serret-Frenet formulae to particle dynamics we make the following identifications:

1. The space curve $\mathcal{C}$ is identified as the path of the particle.

2. The arc-length parameter s is considered to be a function of t.

In particular, we note that s may be identified as the distance traveled along the curve $\mathcal{C}$ from a given reference point.

With the identifications in mind, the position vector $\mathbf{r}$ of the particle can be given the equivalent functional representations

$$\mathbf{r} = x\mathbf{E}_x + y\mathbf{E}_y + z\mathbf{E}_z = \mathbf{r}(t) = \hat{\mathbf{r}}\left(s(t)\right).$$

Using these representations, we obtain for the velocity vector $\mathbf{v}$ of the particle,

$$\mathbf{v} = \dot{x}\mathbf{E}_x + \dot{y}\mathbf{E}_y + \dot{z}\mathbf{E}_z = \dot{\mathbf{r}}(t) = \frac{d\mathbf{r}}{ds}\frac{ds}{dt} = \frac{ds}{dt}\mathbf{e}_t .$$

It is important to see here that

$$\mathbf{v} = \frac{ds}{dt}\mathbf{e}_t = v\mathbf{e}_t .$$

Since the speed v is greater than or equal to zero, $\mathbf{e}_t$ is the unit vector in the direction of $\mathbf{v}$. Similarly, for the acceleration vector $\mathbf{a}$ we obtain the expression

$$\mathbf{a} = \dot{\mathbf{v}} = \frac{d}{dt}\left(\frac{ds}{dt}\mathbf{e}_t\right) = \frac{d^2s}{dt^2}\mathbf{e}_t + \frac{ds}{dt}\frac{d\mathbf{e}_t}{dt} = \frac{d^2s}{dt^2}\mathbf{e}_t + \frac{ds}{dt}\frac{d\mathbf{e}_t}{ds}\frac{ds}{dt} .$$

Recalling the definitions of the principal normal vector $\mathbf{e}_n$ and the speed $v = \dot{s}$, we obtain the final desired expression for $\mathbf{a}$:

$$\mathbf{a} = \dot{\mathbf{v}} = \frac{d^2s}{dt^2}\mathbf{e}_t + \kappa\left(\frac{ds}{dt}\right)^2 \mathbf{e}_n = \frac{dv}{dt}\mathbf{e}_t + \kappa v^2 \mathbf{e}_n .$$

This remarkable result states that the acceleration vector of the particle lies entirely in the osculating plane (see Figure 3.5).

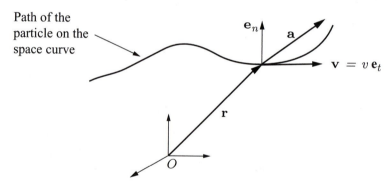

Path of the particle on the space curve

FIGURE 3.5. The velocity and acceleration vectors of a particle moving on a space curve

We note in passing that the distance traveled by the particle along its path may be determined from the vector $\mathbf{v}$. To see this, we first recall that

$$\mathbf{v} \cdot \mathbf{v} = \frac{d\mathbf{r}}{dt} \cdot \frac{d\mathbf{r}}{dt} = \left(\frac{ds}{dt}\right)^2 .$$

This implies that

$$s - s_0 = \int_{s_0}^{s} ds = \int_{t_0}^{t} \sqrt{\mathbf{v}(\mu) \cdot \mathbf{v}(\mu)}\,d\mu ,$$

where s_0 denotes the value of s when $t = t_0$.

For a particle of mass m, Newton's second law states that

$$\mathbf{F} = m\mathbf{a},$$

where $\mathbf{F}$ is the resultant external force acting on the particle. Recalling that, for each s, the set of vectors $\{\mathbf{e}_t, \mathbf{e}_n, \mathbf{e}_b\}$ forms a basis for $\mathcal{E}^3$, we may write

$$\mathbf{F} = F_t\mathbf{e}_t + F_n\mathbf{e}_n + F_b\mathbf{e}_b,$$

where

$$
\begin{aligned}
F_t &= \hat{F}_t(s) = \mathbf{F} \cdot \hat{\mathbf{e}}_t(s), \\
F_n &= \hat{F}_n(s) = \mathbf{F} \cdot \hat{\mathbf{e}}_n(s), \\
F_b &= \hat{F}_b(s) = \mathbf{F} \cdot \hat{\mathbf{e}}_b(s).
\end{aligned}
$$

Using these results, we find that

$$
\begin{aligned}
(\mathbf{F} = m\mathbf{a}) \cdot \mathbf{e}_t &: \quad F_t = m\frac{d^2s}{dt^2}, \\
(\mathbf{F} = m\mathbf{a}) \cdot \mathbf{e}_n &: \quad F_n = m\kappa\left(\frac{ds}{dt}\right)^2, \\
(\mathbf{F} = m\mathbf{a}) \cdot \mathbf{e}_b &: \quad F_b = 0.
\end{aligned}
$$

In certain cases, these 3 equations are completely uncoupled and allow a problem in particle dynamics to be easily solved. We note that the previous equations also imply that $\mathbf{F}$ lies entirely in the osculating plane.

3.5 A Particle Moving on a Fixed Curve Under Gravity

As shown in Figure 3.6, we consider a particle of mass m moving on a smooth plane curve $y = f(x) = -x^2$. A vertical gravitational force $-mg\mathbf{E}_y$ acts on the particle. We seek to determine the differential equation governing the motion of the particle and the force exerted by the curve on the particle. In addition, we examine the conditions on the motion of the particle that result in its losing contact with the curve.

3.5.1 Kinematics

We first consider the Serret-Frenet triad for this plane curve. From the results in Section 3.1, we find that the arc-length parameter s of the curve

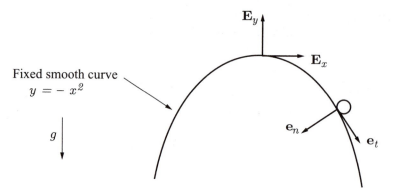

FIGURE 3.6. A particle moving on a curve under gravity

as a function of x is

$$
s \;=\; s(x) = \int_0^x \sqrt{1 + 4u^2}\,du + s(0)
$$

$$
\;=\; \frac{x}{2}\sqrt{1 + 4x^2} + \frac{1}{4}\sinh^{-1}(2x) + s(0)\,.
$$

Here, we have taken the arbitrary constant $x_0 = 0$. One can also determine the radius of curvature:

$$
\rho = \frac{1}{2}\left(\sqrt{1 + 4x^2}\right)^3 .
$$

The Serret-Frenet triad as a function of x can be calculated:

$$
\mathbf{e}_t \;=\; \frac{1}{\sqrt{1 + 4x^2}}\,\left(\mathbf{E}_x - 2x\mathbf{E}_y\right)\,,
$$

$$
\mathbf{e}_n \;=\; \frac{-1}{\sqrt{1 + 4x^2}}\,\left(\mathbf{E}_y + 2x\mathbf{E}_x\right)\,,
$$

$$
\mathbf{e}_b \;=\; -\mathbf{E}_z\,.
$$

The kinematics of the particle are given by the formulae in Section 4:

$$
\mathbf{a} = \dot{\mathbf{v}} = \frac{d^2 s}{dt^2}\mathbf{e}_t + \kappa\left(\frac{ds}{dt}\right)^2 \mathbf{e}_n\,.
$$

3.5.2   Forces

As shown in Figure 3.7, we next consider a free-body diagram of the particle. The forces acting on the particle are due to the gravitational force and the normal force $\mathbf{N} = N_n\mathbf{e}_n + N_b\mathbf{e}_b$, which is the force that the curve exerts on the particle:

$$
\mathbf{F} = -mg\mathbf{E}_y + N_n\mathbf{e}_n + N_b\mathbf{e}_b\,.
$$

FIGURE 3.7. Free-body diagram of the particle

3.5.3   Balance Law

Taking the components of $\mathbf{F} = m\mathbf{a}$ with respect to the Serret-Frenet triad, 3 scalar equations are obtained:

$$m\ddot{s} = \frac{2xmg}{\sqrt{1 + 4x^2}} \, , \qquad N_n = \frac{m\dot{s}^2}{\rho} - \frac{mg}{\sqrt{1 + 4x^2}} \, , \qquad N_b = 0 \, .$$

3.5.4   Analysis

The last two of the above equations determine the normal force $\mathbf{N}$:

$$\mathbf{N} = \left(\frac{mv^2}{\rho} - \frac{mg}{\sqrt{1 + 4x^2}} \right) \mathbf{e}_n \, .$$

The first equation above determines the motion of the particle on the curve:

$$m\dot{v} = \frac{2xmg}{\sqrt{1 + 4x^2}} \, .$$

To proceed to describe this equation as a differential equation for $x(t)$, we note that

$$v = \frac{ds}{dt} = \frac{ds\,dx}{dx\,dt} = \frac{dx}{dt}\sqrt{1 + 4x^2} \, ,$$

$$\frac{dv}{dt} = \frac{d^2 s}{dt^2} = \frac{d^2 s}{dx^2}\left(\frac{dx}{dt}\right)^2 + \frac{ds\,d^2 x}{dx\,dt^2}$$

$$= \frac{4x}{\sqrt{1 + 4x^2}}\left(\frac{dx}{dt}\right)^2 + \sqrt{1 + 4x^2}\,\frac{d^2 x}{dt^2} \, .$$

You should note that $v \neq dx/dt$. For the example at hand, we substitute these results into the differential equation for $\frac{dv}{dt}$. After some rearranging, we obtain the desired ordinary differential equation:

$$\frac{d^2 x}{dt^2} = \frac{1}{1 + 4x^2}\left(2xg - 4x\left(\frac{dx}{dt}\right)^2 \right) \, .$$

Given the initial conditions $x(t_0)$ and $\dot{x}(t_0)$, the solution $x(t)$ of this equation can then be used to determine the motion $\mathbf{r}(t) = x(t)\mathbf{E}_x - x^2(t)\mathbf{E}_y$ of the particle, and the force $\mathbf{N}(t)$ exerted by the curve on the particle.

The ordinary differential equation governing $x(t)$ is formidable. It is non-linear and well beyond the scope of an undergraduate engineering dynamics course to solve. Instead, one is content with finding the speed v as a function of x. To this end, one uses the identity

$$a_t = \mathbf{a} \cdot \mathbf{e}_t = \frac{dv}{dt} = \frac{dv}{dx}\frac{dx}{dt} = \frac{v}{\sqrt{1+4x^2}}\frac{dv}{dx} \, .$$

It follows that

$$v^2(x) = v^2(x_0) + 2\int_{x_0}^{x} a_t(u)\sqrt{1+4u^2}\, du \, .$$

For the example at hand,

$$a_t(x) = \frac{2xg}{\sqrt{1+4x^2}} \, ,$$

and hence,

$$v^2(x) = v^2(x_0) + \int_{x_0}^{x} 4\,u\,g\,du = v^2(x_0) + 2g\left(x^2 - x_0^2\right) \, .$$

Here, x_0 and $v(x_0)$ are given initial conditions. Since $y = -x^2$, some of you may notice that this is none other than a conservation of energy result (cf. the conclusion of Section 6 of Chapter 5, where this problem is revisited).

We are now in a position to establish a criterion for the particle leaving the curve. One can use the previous equation and the expression for ρ to calculate the force $\mathbf{N}$ as a function of x:

$$\mathbf{N} = \left(\frac{mv^2}{\rho} - \frac{mg}{\sqrt{1+4x^2}}\right)\mathbf{e}_n = \left(\frac{2mv^2(x_0) - mg(1+4x_0^2)}{\left(\sqrt{1+4x^2}\right)^3}\right)\mathbf{e}_n \, .$$

Notice that if $v(x_0)$ is sufficiently large, then $\mathbf{N} \cdot \mathbf{e}_n \geq 0$ and the particle will not remain on the curve. Specifically, if

$$v(x_0) \geq \sqrt{\frac{g}{2}(1+4x_0^2)} \, ,$$

then the particle immediately loses contact with the curve.

3.6 Summary

This chapter established the machinery needed to examine the dynamics of particles moving in a general manner in three-dimensional space. To this end, some results pertaining to curves in three-dimensional space were presented.

For a given space curve, the three vectors introduced in this chapter, the Serret-Frenet basis vectors $\{\mathbf{e}_t, \mathbf{e}_n, \mathbf{e}_b\}$, form a right-handed orthonormal basis at each point of the curve. The three vectors are defined by

$$\mathbf{e}_t = \frac{d\mathbf{r}}{ds}, \quad \kappa\mathbf{e}_n = \frac{d^2\mathbf{r}}{ds^2}, \quad \mathbf{e}_b = \mathbf{e}_t \times \mathbf{e}_n .$$

The rate of change of these vectors as the arc-length parameter s of the curve varies is given by the Serret-Frenet relations:

$$\frac{d\mathbf{e}_t}{ds} = \kappa\mathbf{e}_n , \quad \frac{d\mathbf{e}_n}{ds} = -\kappa\mathbf{e}_t + \tau\mathbf{e}_b , \quad \frac{d\mathbf{e}_b}{ds} = -\tau\mathbf{e}_n .$$

These relations and the Serret-Frenet basis vectors were illustrated for the case of a plane curve, a straight line, a circle, a space curve parametrized by x, and a circular helix. For three of these examples, it was convenient to describe the Serret-Frenet basis vectors, torsion τ, and curvature κ as functions of x rather than s. In a similar manner, the variable θ was used for the circle and circular helix.

To use these results in particle dynamics, the path of the particle is identified with a space curve. Then, we showed that the following representations held:

$$\mathbf{v} = \dot{s}\frac{d\mathbf{r}}{ds} = \dot{s}\mathbf{e}_t = v\mathbf{e}_t ,$$

$$\mathbf{a} = \dot{s}^2\frac{d^2\mathbf{r}}{ds^2} + \ddot{s}\frac{d\mathbf{r}}{ds} = \kappa\dot{s}^2\mathbf{e}_n + \ddot{s}\mathbf{e}_t = \kappa v^2\mathbf{e}_n + \dot{v}\mathbf{e}_t .$$

In these equations, $v = \dot{s}$. Using the Serret-Frenet basis vectors, the balance of linear momentum can be described by the following three equations:

$$F_t = \mathbf{F} \cdot \mathbf{e}_t = m\dot{v} , \quad F_n = \mathbf{F} \cdot \mathbf{e}_n = m\kappa v^2 , \quad F_b = \mathbf{F} \cdot \mathbf{e}_b = 0 .$$

These equations were illustrated using the example of a particle moving on a smooth fixed curve.

3.7   Exercises

The following short exercises are intended to assist you in reviewing Chapter 3.

3.1   For the space curve $\mathbf{r} = x\mathbf{E}_x + ax\mathbf{E}_y$, show that $\mathbf{e}_t = \frac{1}{\sqrt{1+a^2}}(\mathbf{E}_x + a\mathbf{E}_y)$ and $\kappa = 0$. In addition, show that $\mathbf{e}_n$ is any vector perpendicular to $\mathbf{e}_t$, for example, $\mathbf{e}_n = \frac{1}{\sqrt{1+a^2}}(-a\mathbf{E}_x + \mathbf{E}_y)$. Why is the torsion τ of this curve 0?

3.2 Calculate the Serret-Frenet basis vectors, for the space curve $\mathbf{r} = x\mathbf{E}_x + \frac{x^3}{3}\mathbf{E}_y$. It is convenient to describe these vectors as functions of x. In addition, show that the arc-length parameter s as a function of x is given by

$$s = s(x) = \int_{x_0}^{x} \sqrt{1 + u^4}\,du + s(x_0)\,.$$

3.3 Consider the space curve $\mathbf{r} = x\mathbf{E}_x + \frac{x^3}{3}\mathbf{E}_y + \frac{x^2}{2}\mathbf{E}_z$. Calculate the Serret-Frenet basis vectors for this curve. Again, it is convenient to describe these vectors as functions of x. In addition, show that the arc-length parameter s as a function of x is given by

$$s = s(x) = \int_{x_0}^{x} \sqrt{1 + u^4 + u^2}\,du + s(x_0)\,.$$

3.4 Consider the plane circle $\mathbf{r} = 10\mathbf{e}_r$. Show that $\mathbf{e}_t = \mathbf{e}_\theta$, $\mathbf{e}_n = -\mathbf{e}_r$, $\kappa = 0.1$, $\rho = 10$, $\tau = 0$, and $\mathbf{e}_b = \mathbf{E}_z$.

3.5 Calculate the Serret-Frenet basis vectors for the circular helix $\mathbf{r} = 10\mathbf{e}_r + 10\theta\mathbf{E}_z$. In addition, show that $s(\theta) = 10\sqrt{2}\,(\theta - \theta_0) + s(\theta_0)$ and $\kappa = \tau = \frac{1}{20}$.

3.6 Starting from $\mathbf{r} = \mathbf{r}(s(t))$, show that $\mathbf{v} = v\mathbf{e}_t$ and $\mathbf{a} = \dot{v}\mathbf{e}_t + \kappa v^2 \mathbf{e}_n$.

3.7 A section of track of a roller coaster can be defined using the space curve $\mathbf{r} = x\mathbf{E}_x + f(x)\mathbf{E}_x$. The measurement system used to determine the speed of a trolley moving on this track measures $x(t)$ rather than $s(t)$. Consequently, in order to establish and verify the equations of motion of the trolley, it is desirable to know $\dot{x}$ and $\ddot{x}$ in terms of $\dot{s}$ and $\ddot{s}$. Starting from the definition $\dot{s} = \|\mathbf{v}\|$, show that

$$\dot{s} = \sqrt{1 + \left(\frac{df}{dx}\right)^2}\,\dot{x}\,.$$

Using this result and the chain rule, show that

$$\ddot{s} = \frac{1}{\sqrt{1 + \left(\frac{df}{dx}\right)^2}}\left(\frac{df}{dx}\frac{d^2 f}{dx^2}\dot{x}^2 + \left(1 + \left(\frac{df}{dx}\right)^2\right)\ddot{x}\right)\,.$$

3.8 Using the results of Exercise 3.7 and the expressions for the Serret-Frenet basis vectors recorded in Section 3.1, write out the equations governing the motion of the trolley. You should assume that a gravitational force $-mg\mathbf{E}_y$ acts on the trolley while it is moving on the track, and model the trolley as a particle of mass m.

3.9 Consider a particle of mass m moving on a circular helix $\mathbf{r} = R\mathbf{e}_R + \alpha R\theta\mathbf{E}_z$. A gravitational force $-mg\mathbf{E}_z$ acts on the particle. Show that

$$s(\theta) = R\sqrt{1+\alpha^2}\,(\theta - \theta_0) + s(\theta_0)\,, \quad \ddot{s} = R\sqrt{1+\alpha^2}\ddot{\theta}\,.$$

Using $\mathbf{F} = m\mathbf{a}$ and the results of Section 3.4, show that

$$\mathbf{F} \cdot \mathbf{e}_t = -\frac{mg\alpha}{\sqrt{1+\alpha^2}} = mR\sqrt{1+\alpha^2}\ddot{\theta}\,,$$

$$\mathbf{F} \cdot \mathbf{e}_n = N_n = mR\dot{\theta}^2\,,$$

$$\mathbf{F} \cdot \mathbf{e}_b = N_b - \frac{mg}{\sqrt{1+\alpha^2}}\,.$$

Here, $N_n\mathbf{e}_n + N_b\mathbf{e}_b$ is the normal force exerted by the curve on the particle.

3.10 Show that the polar coordinate θ of the particle discussed in Exercise 3.9 is

$$\theta(t) = -\frac{g\alpha}{2R(1+\alpha^2)}\,(t - t_0)^2 + \dot{\theta}_0\,(t - t_0) + \theta_0\,.$$

For various initial conditions, θ_0 and $\dot{\theta}_0$, discuss the motion of the particle on the helix. You should notice the similarities between the results of this exercise and the projectile problem discussed in Chapter 1.

4
Friction Forces and Spring Forces

Topics

Two types of forces are discussed in this chapter: friction forces and spring forces. We start with the former and consider a simple classical experiment. Based on this experiment, general (coordinate-free) expressions for friction forces are obtained. The chapter closes with the corresponding developments for a spring force.

The reason for including a separate chapter on these two types of forces is that I have found that they present the most difficulty to students when they are formulating problems. In particular, many students have the impression that if they do not correctly guess the direction of spring and friction forces in their free-body diagram, then they will get the wrong answer. The coordinate-free formulation of these forces in this chapter bypasses this issue.

4.1 An Experiment on Friction

Most theories of friction forces arise from studies by the French scientist Charles Augustin Coulomb.[1] Here, we review a simple experiment which

---

[1] The most-cited reference to his work is his prize-winning paper [16], which was published in 1785. Accounts of Coulomb's work on friction are contained in Dugas [21] and Heyman [32].

is easily replicated (at least qualitatively) using a blackboard eraser, some weights, and a table.

As shown in Figure 4.1, consider a block of mass m that is initially at rest on a horizontal plane. A force $P\mathbf{E}_x$ acts on the block.

FIGURE 4.1. A block on a rough horizontal plane

Upon increasing P, the following observations can be made:

(a) For small values of P, the block remains at rest.

(b) Beyond a critical value $P = P^*$, the block starts to move.

(c) Once in motion, a constant value of $P = P^{**}$ is required to move the block at a constant speed.

(d) Both P^* and P^{**} are proportional to the magnitude of the normal force $\mathbf{N}$.

Let's now analyze this experiment. We model the block as a particle of mass m. Let

$$\mathbf{r} = x\mathbf{E}_x + y_0\mathbf{E}_y + z_0\mathbf{E}_z \, ,$$

where y_0 and z_0 are constants. We leave it as an exercise to derive expressions for $\mathbf{v}$ and $\mathbf{a}$. The free-body diagram of the particle is shown in Figure 4.2.

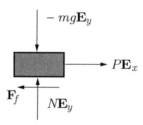

FIGURE 4.2. Free-body diagram of the block

In Figure 4.2, $\mathbf{F}_f = F_{fx}\mathbf{E}_x + F_{fz}\mathbf{E}_z$ is the friction force exerted by the surface on the particle, while $\mathbf{N} = N\mathbf{E}_y$ is the normal (or reaction) force exerted by the surface on the particle.

From $\mathbf{F} = m\mathbf{a}$, we obtain 3 equations:

$$
\begin{aligned}
F_{fx} + P &= m\ddot{x}\,, \\
N - mg &= 0\,, \\
F_{fz} &= 0\,.
\end{aligned}
$$

For the static case, $\ddot{x} = 0$, and from $\mathbf{F} = m\mathbf{a}$ we find that $\mathbf{F}_f = -P\mathbf{E}_x$. As noted previously, as the magnitude of $\mathbf{P}$ is increased beyond a critical value P^*, the block starts to move. The critical value of P^* is proportional to the magnitude of the normal force $\mathbf{N}$. We denote this constant of proportionality by the coefficient of static friction μ_s. Hence, $|F_{fx}| \leq \mu_s|N| = \mu_s mg$ and $P^* = \mu_s mg$. In summary, for the static case, $\mathbf{F}_f = -P\mathbf{E}_x$ until $P = \mu_s mg$. As P increases beyond this value, then the block moves and the friction force is no longer equal to $-P\mathbf{E}_x$.

When the block starts moving, then the friction force opposes the motion. As a result, its direction is opposite to $\mathbf{v}$. Further, as noted above, its magnitude is proportional to the magnitude of $\mathbf{N}$. The constant of proportionality is denoted by μ_d, the coefficient of dynamic friction. Hence, assuming that the block moves to the right: $\mathbf{F}_f = -\mu_d|N|\mathbf{E}_x = -\mu_d mg\mathbf{E}_x$. If the block moves at constant speed, then we find from $\mathbf{F} = m\mathbf{a}$ that $P^{**} = \mu_d mg$.

The coefficients of static friction and dynamic friction depend on the nature of the surfaces of the horizontal surface and the block. They must be determined experimentally.

4.2 Static and Dynamic Coulomb Friction Forces

Among other assumptions, the previous developments assumed that the horizontal surface was flat and stationary. For many problems, our developments are insufficient and need to be generalized in several directions:

(i) Cases where the surface on which the particle lies is curved.

(ii) Cases where the particle is moving on a space curve.

(iii) Cases where the space curve or surface on which the particle moves is in motion.

It is to these cases that we now turn. The theory that we present here is not universally applicable, although it is used extensively in engineering.[2] For example, the friction between the surface of a road and the tires of a vehicle is not, in general, of the Coulomb type, but it is often modeled as such.

---

[2] For further information on the limitations of the Coulomb friction theory, we refer the reader to Rabinowicz [49] and Ruina [54]

To proceed, the position vector and absolute velocity vector of the particle are denoted by $\mathbf{r}$ and $\mathbf{v}$, respectively. If the particle is in motion on a space curve, then the velocity vector of the point of the space curve which is in contact with the particle is denoted by $\mathbf{v}_c$. Similarly, if the particle is in motion on a surface, then the velocity vector of the point of the surface which is in contact with the particle is denoted by $\mathbf{v}_s$.

4.2.1 A Particle on a Surface

Here, we assume that the particle is moving on a surface. Referring to Figure 4.3, at the point P of the surface that is in contact with the particle, we assume that there is a well-defined unit normal vector $\mathbf{n}$. At this point of contact one also has two unit tangent vectors $\mathbf{t}_1$ and $\mathbf{t}_2$. We choose these tangent vectors such that $\{\mathbf{t}_1, \mathbf{t}_2, \mathbf{n}\}$ is a right-handed basis for Euclidean three-space.

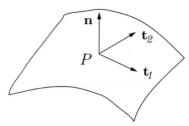

FIGURE 4.3. The two tangent vectors and normal vector at a point P of a surface

It is convenient to consider some examples. First, if the particle is moving on a horizontal plane, then $\mathbf{n} = \mathbf{E}_z$, $\mathbf{t}_1 = \mathbf{E}_x$, and $\mathbf{t}_2 = \mathbf{E}_y$. Another example, which we will examine later, is a particle on the inner surface of a cone $z = r\tan(\alpha)$. Here, α is the angle of inclination of the conical surface. In this case, we have

$$\mathbf{n} = \cos(\alpha)\mathbf{E}_z - \sin(\alpha)\mathbf{e}_r, \quad \mathbf{t}_1 = \cos(\alpha)\mathbf{e}_r + \sin(\alpha)\mathbf{E}_z, \quad \mathbf{t}_2 = \mathbf{e}_\theta.$$

Notice how these vectors have been normalized so as to have unit magnitude.

Recall that $\mathbf{v}_s$ denotes the absolute velocity vector of the point of contact P. Then, the velocity vector of the particle relative to P is

$$\mathbf{v}_{\mathrm{rel}} = \mathbf{v} - \mathbf{v}_s = v_1\mathbf{t}_1 + v_2\mathbf{t}_2.$$

If $\mathbf{v}_{\mathrm{rel}} = \mathbf{0}$, then the particle is said to be stationary relative to the surface. Specifically, we have

$$\text{static friction: } \mathbf{v}_{\mathrm{rel}} = \mathbf{0},$$
$$\text{dynamic friction: } \mathbf{v}_{\mathrm{rel}} \neq \mathbf{0}.$$

The force exerted by the surface on the particle is composed of two parts: the normal (or reaction) force $\mathbf{N}$ and the friction force $\mathbf{F}_f$. For both the static and dynamic cases,

$$\mathbf{N} = N\mathbf{n}\,.$$

Further, N is indeterminate. It can be found only from $\mathbf{F} = m\mathbf{a}$. The static friction force is

$$\mathbf{F}_f = F_{f1}\mathbf{t}_1 + F_{f2}\mathbf{t}_2\,,$$

where F_{f1} and F_{f2} are also indeterminate. The amount of static friction available is limited by the coefficient of static friction:

$$\|\mathbf{F}_f\| \leq \mu_s\|\mathbf{N}\|\,.$$

If this criterion fails, then the particle will move relative to the surface. The friction force in this case is dynamic:

$$\mathbf{F}_f = -\mu_d\|\mathbf{N}\|\frac{\mathbf{v}_{\text{rel}}}{\|\mathbf{v}_{\text{rel}}\|}\,.$$

You should notice that this force opposes the motion of the particle relative to the surface.

4.2.2 A Particle on a Space Curve

Here, we assume that the particle is moving on a curve. At the point P of the curve that is in contact with the particle, we assume that there is a well-defined unit tangent vector $\mathbf{e}_t$ (see Figure 4.4). At this point of contact, one also has two unit normal vectors $\mathbf{n}_1$ and $\mathbf{n}_2$. We choose these tangent vectors such that $\{\mathbf{n}_1, \mathbf{n}_2, \mathbf{e}_t\}$ is a right-handed basis for Euclidean three-space. The vectors $\mathbf{n}_1$ and $\mathbf{n}_2$ lie in the plane spanned by $\mathbf{e}_n$ and $\mathbf{e}_b$. For instance, if the particle is moving on a horizontal line, then we can choose $\mathbf{e}_t = \mathbf{E}_x$, $\mathbf{n}_1 = \mathbf{E}_y$, and $\mathbf{n}_2 = \mathbf{E}_z$.

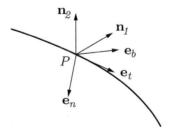

FIGURE 4.4. The tangent vector and two normal vectors at a point P of a curve

Recall that $\mathbf{v}_c$ denotes the absolute velocity vector of the point of contact P. Then, the velocity vector of the particle relative to P is

$$\mathbf{v}_{\text{rel}} = \mathbf{v} - \mathbf{v}_c = v_t\mathbf{e}_t\,.$$

If $\mathbf{v}_{\mathrm{rel}} = \mathbf{0}$, then the particle is said to be stationary relative to the curve. Specifically, we again have that

$$\text{static friction: } \mathbf{v}_{\mathrm{rel}} = \mathbf{0},$$
$$\text{dynamic friction: } \mathbf{v}_{\mathrm{rel}} \neq \mathbf{0}.$$

The force exerted by the curve on the particle is composed of two parts: the normal (or reaction) force $\mathbf{N}$ and the friction force $\mathbf{F}_f$. For both the static and dynamic cases,

$$\mathbf{N} = N_1 \mathbf{n}_1 + N_2 \mathbf{n}_2 \, .$$

Further, N_1 and N_2 are indeterminate. They can be found only from $\mathbf{F} = m\mathbf{a}$. The static friction force is

$$\mathbf{F}_f = F_f \mathbf{e}_t \, ,$$

where F_f is also indeterminate. The amount of static friction available is limited by the coefficient of static friction:

$$\|\mathbf{F}_f\| \leq \mu_s \|\mathbf{N}\| \, .$$

Specifically,

$$|F_f| \leq \mu_s \sqrt{N_1^2 + N_2^2} \, .$$

If this criterion fails, then the particle will move relative to the curve. The friction force in this case is dynamic:

$$\begin{aligned}
\mathbf{F}_f &= -\mu_d \|\mathbf{N}\| \frac{\mathbf{v}_{\mathrm{rel}}}{\|\mathbf{v}_{\mathrm{rel}}\|} \\
&= -\mu_d \sqrt{N_1^2 + N_2^2} \, \frac{v_t}{|v_t|} \mathbf{e}_t \, .
\end{aligned}$$

You should notice that this force opposes the motion of the particle relative to the curve.

4.3   Some Comments on Friction Forces

It is important to note that in the static friction case, the friction force $\mathbf{F}_f$ and the normal force $\mathbf{N}$ are unknown. However, since the motion of the surface/curve is known, the motion of the particle is also known. Hence, $\mathbf{F} = m\mathbf{a}$ provides three equations to determine these forces.

An error many students make is setting the static friction force F_{fx}, say, equal to $\mu_s N$ or, worse, $\mu_s mg$. Setting the static friction force equal to one of its maximum values is generally not valid.

The careful reader will have noted that we are using $\mathbf{e}_t$ to denote the tangent vector to a moving curve. However, our previous developments in

Chapter 3 were limited to a fixed curve. As mentioned there, they can be extended to a moving curve. Specifically, let $\mathbf{p}$ denote the position vector to any point on a moving space curve. This position vector depends on the arc-length parameter and time:

$$\mathbf{p} = \tilde{\mathbf{p}}(s, t) \,.$$

One now has the following definitions of the Serret-Frenet triad:

$$\mathbf{e}_t = \tilde{\mathbf{e}}_t(s, t) = \frac{\partial \mathbf{p}}{\partial s} \,, \quad \kappa \mathbf{e}_n = \frac{\partial \mathbf{e}_t}{\partial s} \,, \quad \mathbf{e}_b = \mathbf{e}_t \times \mathbf{e}_n \,.$$

It should be obvious that these definitions parallel the developments for a fixed curve, except that the derivatives are partial derivatives. In essence one is *freezing* time and evaluating the Serret-Frenet triad for the *frozen* curve. Another point of interest is that the velocity $\mathbf{v}_c$ discussed earlier is equal to $\frac{\partial \mathbf{p}}{\partial t}$. It is interesting to note that these developments are used in theories of rods (see, for example, Antman [1] or Love [37]).

4.4   A Particle on a Rough Moving Plane

To illustrate the previous developments, we consider the problem of a particle moving a rough horizontal plane as shown in Figure 4.5. Every point on this plane is assumed to be moving with a velocity $\mathbf{v}_s = v_s \mathbf{E}_z$. A vertical gravitational force also acts on the particle.

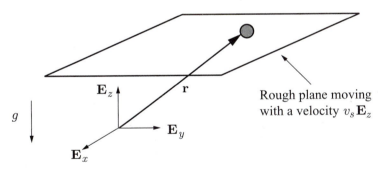

FIGURE 4.5. A particle moving on a horizontal plane

For this problem,

$$\mathbf{t}_1 = \mathbf{E}_x \,, \quad \mathbf{t}_2 = \mathbf{E}_y \,, \quad \mathbf{n} = \mathbf{E}_z \,.$$

Further, the velocity vector of the particle is

$$\mathbf{v} = \dot{x}\mathbf{E}_x + \dot{y}\mathbf{E}_y + v_s\mathbf{E}_z \,.$$

The relative velocity vector of the particle is

$$\mathbf{v}_{\text{rel}} = \mathbf{v} - \mathbf{v}_s = \dot{x}\mathbf{E}_x + \dot{y}\mathbf{E}_y\,.$$

We leave it as an exercise to draw the free-body diagrams of the particle for the static and dynamic cases.

Turning to the results for the static case, we have

$$\mathbf{F} = N\mathbf{E}_z + F_{fx}\mathbf{E}_x + F_{fy}\mathbf{E}_y - mg\mathbf{E}_z = m\mathbf{a} = m\dot{v}_s\mathbf{E}_z\,.$$

It follows from these equations that

$$\mathbf{F}_f = \mathbf{0}\,, \quad \mathbf{N} = m(g + \dot{v}_s)\mathbf{E}_z\,.$$

As expected, the friction force is zero in this case.

Turning to the results for the dynamic case, we have

$$\mathbf{F} = N\mathbf{E}_z - \mu_d|N|\frac{\mathbf{v}_{\text{rel}}}{\|\mathbf{v}_{\text{rel}}\|} - mg\mathbf{E}_z = m\mathbf{a}\,.$$

From these 3 scalar equations, we find that the normal force is

$$\mathbf{N} = m(g + \dot{v}_s)\mathbf{E}_z\,.$$

In addition, 2 ordinary differential equations of the motion of the particle relative to the surface are obtained:

$$m\ddot{x} = -\mu_d|m(g + \dot{v}_s)|\frac{\dot{x}}{\sqrt{\dot{x}^2 + \dot{y}^2}}\,, \quad m\ddot{y} = -\mu_d|m(g + \dot{v}_s)|\frac{\dot{y}}{\sqrt{\dot{x}^2 + \dot{y}^2}}\,.$$

As expected, these differential equations are valid only when $\|\mathbf{v}_{\text{rel}}\|$ is non-zero.

4.5   Hooke's Law and Linear Springs

The classical result on linear springs is due to a contemporary of Isaac Newton, Robert Hooke (1635–1703), who announced his result in the form of the anagram *ceiiinosssttuu* in 1660. He later published his result in his work *Lectures de Potentia restitutiva, or, Of spring explaining the power of springing bodies* in 1678:

Ut tensio sic vis

which tranlates to "the power of any spring is in the same proportion with the tension thereof."[3] His result is often known as Hooke's law.

---

[3]From the *Historical Introduction* to Love [37]. In modern terminology, *power* is force and *tension* is extension.

As with Coulomb's work on friction, Hooke's law is based on experimental evidence. It is not valid in all situations. For instance, it implies that it is possible to extend a spring as much as desired without the spring breaking, which is patently not true. However, for many applications, where the change in the spring's length is small, it is a valid and extremely useful observation.

Referring to Figure 4.6, we wish to develop a general result for the force exerted by a spring on a particle when the motion of the particle changes the length of the spring. We confine our attention to a linear spring. That is, we assume that Hooke's law is valid.

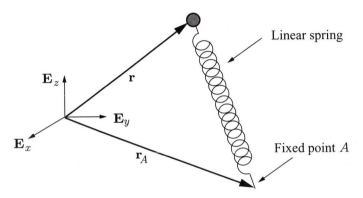

FIGURE 4.6. A spring attached to a mass particle

As shown in Figure 4.6, we consider a massless spring of stiffness K and unstretched length L. One end of the spring is attached to a point A which has a position vector $\mathbf{r}_A$. Its other end is attached to a mass particle whose position vector is $\mathbf{r}$.

The force generated by the spring is assumed to be linearly proportional to its extension/compression. Here, the change in length of the spring is

$$\|\mathbf{r} - \mathbf{r}_A\| - L \,.$$

If this number is positive, then the spring is extended. The magnitude of the spring force $\mathbf{F}_s$ is

$$\|\mathbf{F}_s\| = |K\left(\|\mathbf{r} - \mathbf{r}_A\| - L\right)| \,.$$

This is a statement of Hooke's law.

It remains to determine the direction of $\mathbf{F}_s$. First, suppose the spring is extended. Then the force $\mathbf{F}_s$ will attempt to pull the particle towards A. In other words, its direction is

$$-\frac{\mathbf{r} - \mathbf{r}_A}{\|\mathbf{r} - \mathbf{r}_A\|} \,.$$

Combining this observation with the result on the magnitude of the spring force, and noting that the extension is positive, we arrive at the following expression:

$$\mathbf{F}_s = -K\left(\|\mathbf{r} - \mathbf{r}_A\| - L\right) \frac{\mathbf{r} - \mathbf{r}_A}{\|\mathbf{r} - \mathbf{r}_A\|} .$$

On the other hand, if the spring is compressed, then this force will attempt to push the particle away from A. As a result its direction will be

$$\frac{\mathbf{r} - \mathbf{r}_A}{\|\mathbf{r} - \mathbf{r}_A\|} .$$

Since the change in length of the spring force is negative, we find that the magnitude of the spring force in this case is

$$\|\mathbf{F}_s\| = -K\left(\|\mathbf{r} - \mathbf{r}_A\| - L\right) .$$

Consequently, the spring force when the spring is compressed is

$$\mathbf{F}_s = -K\left(\|\mathbf{r} - \mathbf{r}_A\| - L\right) \frac{\mathbf{r} - \mathbf{r}_A}{\|\mathbf{r} - \mathbf{r}_A\|} .$$

It should be clear that the final expressions for $\mathbf{F}_s$ we have obtained for the extended and compressed springs are identical. In summary, then for a spring of stiffness K and unstretched length L, the force exerted by the spring on the particle is

$$\mathbf{F}_s = -K\left(\|\mathbf{r} - \mathbf{r}_A\| - L\right) \frac{\mathbf{r} - \mathbf{r}_A}{\|\mathbf{r} - \mathbf{r}_A\|} .$$

If for a specific problem one can choose the point A to be the origin, then this expression simplifies considerably to

$$\mathbf{F}_s = -K\left(\|\mathbf{r}\| - L\right) \frac{\mathbf{r}}{\|\mathbf{r}\|} .$$

4.6   A Particle on a Rough Spinning Cone

We now consider the dynamics of a particle on a rough circular cone $z = r\tan(\alpha)$. The cone is rotating about its axis of symmetry with an angular speed $\Omega = \Omega(t)$. As shown in Figure 4.7, the particle is attached to the apex of the cone by a spring of stiffness K and unstretched length L.

In what follows, we seek to determine the differential equations governing the motion of the particle and, in the event that the particle is not moving, the force exerted by the surface on the particle. The problem presented here is formidable; it has spring forces, friction, and a nontrivial surface. However, various special cases of this system arise in many problems in mechanics. For instance, by setting $\alpha = 0$ or 90 degrees, one has the problem of a particle on a cylinder or horizontal plane, respectively.

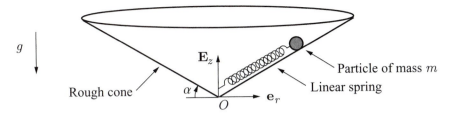

FIGURE 4.7. A particle moving on a rough spinning cone

4.6.1 Kinematics

We choose our origin to coincide with the fixed apex of the cone. Neglecting the thickness of the spring, this point also coincides with the point of attachment of the spring to the apex of the cone. It is convenient to use a cylindrical polar coordinate system to describe the kinematics of this problem:

$$\mathbf{r} = r\mathbf{e}_r + r\tan(\alpha)\mathbf{E}_z .$$

Differentiating this position vector, we find that

$$\mathbf{v} = \dot{r}\left(\mathbf{e}_r + \tan(\alpha)\mathbf{E}_z\right) + r\dot{\theta}\mathbf{e}_\theta ,$$
$$\mathbf{a} = \ddot{r}\left(\mathbf{e}_r + \tan(\alpha)\mathbf{E}_z\right) + \left(r\ddot{\theta} + 2\dot{r}\dot{\theta}\right)\mathbf{e}_\theta - r\dot{\theta}^2\mathbf{e}_r .$$

The position vector of the point of contact P of the particle with the cone is the same as that for the particle. It follows that, at P, the normal and tangential vectors are

$$\mathbf{n} = \cos(\alpha)\mathbf{E}_z - \sin(\alpha)\mathbf{e}_r , \quad \mathbf{t}_1 = \cos(\alpha)\mathbf{e}_r + \sin(\alpha)\mathbf{E}_z , \quad \mathbf{t}_2 = \mathbf{e}_\theta .$$

Turning to the velocity vector of the point P, it is easily seen that this vector is

$$\mathbf{v}_s = r\Omega\mathbf{e}_\theta .$$

Consequently,

$$\mathbf{v}_{\rm rel} = \dot{r}\left(\mathbf{e}_r + \tan(\alpha)\mathbf{E}_z\right) + r(\dot{\theta} - \Omega)\mathbf{e}_\theta .$$

You should notice that, in general, $\dot{\theta} \neq \Omega$. However, if $\mathbf{v}_{\rm rel}$ is zero, then $\mathbf{a}$ simplifies considerably.

4.6.2 Forces

The free-body diagram for this problem is shown in Figure 4.8. Since we have chosen the origin to be the point of attachment of the spring, the spring force has the representation

$$\mathbf{F}_s = -K\left(\|\mathbf{r}\| - L\right)\frac{\mathbf{r}}{\|\mathbf{r}\|} .$$

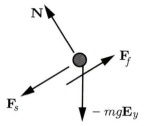

FIGURE 4.8. Free-body diagram of the particle

Furthermore, the normal and friction forces acting on the particle are

$$\mathbf{N} = N\mathbf{n}, \quad \mathbf{F}_f = F_{f1}\mathbf{t}_1 + F_{f\theta}\mathbf{e}_\theta .$$

The resultant force acting on the particle is

$$\mathbf{F} = N\mathbf{n} + F_{f1}\mathbf{t}_1 + F_{f\theta}\mathbf{e}_\theta - mg\mathbf{E}_z - K\left(\|\mathbf{r}\| - L\right)\frac{\mathbf{r}}{\|\mathbf{r}\|} .$$

4.6.3 Balance Law

It is convenient to take the $\mathbf{t}_1$, $\mathbf{e}_\theta$, and $\mathbf{n}$ components of $\mathbf{F} = m\mathbf{a}$:

$$F_{f1} - mg\sin(\alpha) - \frac{K\left(\|\mathbf{r}\| - L\right)r}{\|\mathbf{r}\|\cos(\alpha)} = m\left(\frac{\ddot{r}}{\cos(\alpha)} - r\dot{\theta}^2\cos(\alpha)\right),$$

$$F_{f\theta} = mr\ddot{\theta} + 2m\dot{r}\dot{\theta},$$

$$N - mg\cos(\alpha) = mr\dot{\theta}^2\sin(\alpha).$$

We have omitted several algebraic details that were used to arrive at the final form of these equations.

4.6.4 Analysis

From the 3 equations above, we find that the normal force $\mathbf{N}$ is

$$\mathbf{N} = mg\cos(\alpha)\mathbf{n} + mr\dot{\theta}^2\sin(\alpha)\mathbf{n} .$$

If the particle is not moving relative to the surface, then $\dot{\theta} = \Omega$, $r = r_0$, and this expression becomes

$$\mathbf{N} = mg\cos(\alpha)\mathbf{n} + mr_0\Omega^2\sin(\alpha)\mathbf{n} .$$

In both cases, this force always points in the direction of $\mathbf{n}$, so the particle does not lose contact with the surface.

Turning to the static friction case, we find from the remaining two equations that the friction force is

$$\mathbf{F}_f = \left(\frac{K\left(\|\mathbf{r}_0\| - L\right)r_0}{\|\mathbf{r}_0\|\cos(\alpha)} + mg\sin(\alpha) - mr_0\Omega^2\cos(\alpha)\right)\mathbf{t}_1 + mr_0\dot{\Omega}\mathbf{e}_\theta ,$$

where $\mathbf{r}_0 = r_0 \mathbf{e}_r + r_0 \tan(\alpha)\mathbf{E}_z$. This friction force is limited by the static friction criterion:

$$\|\mathbf{F}_f\| \le \mu_s \|\mathbf{N}\| = \mu_s |mg \cos(\alpha) + mr_0 \Omega^2 \sin(\alpha)| .$$

When this criterion fails, then the particle begins to move relative to the surface. Notice that if Ω is constant, then the static friction force has no component in the $\mathbf{e}_\theta$ direction.

When the particle is in motion relative to the surface, the friction force is

$$\begin{aligned}
\mathbf{F}_f &= -\mu_d \|\mathbf{N}\| \frac{\mathbf{v}_{\mathrm{rel}}}{\|\mathbf{v}_{\mathrm{rel}}\|} \\
&= -\mu_d |mg \cos(\alpha) + mr\dot{\theta}^2 \sin(\alpha)| \frac{\mathbf{v}_{\mathrm{rel}}}{\|\mathbf{v}_{\mathrm{rel}}\|} ,
\end{aligned}$$

where

$$\mathbf{v}_{\mathrm{rel}} = \dot{r}(\mathbf{e}_r + \tan(\alpha)\mathbf{E}_z) + r(\dot{\theta} - \Omega)\mathbf{e}_\theta .$$

Substituting for the friction force in the 2 equations obtained from $\mathbf{F} = m\mathbf{a}$,

$$\begin{aligned}
F_{f1} - mg\sin(\alpha) - \frac{K(\|\mathbf{r}\| - L)r}{\|\mathbf{r}\|\cos(\alpha)} &= m\left(\frac{\ddot{r}}{\cos(\alpha)} - r\dot{\theta}^2 \cos(\alpha)\right) , \\
F_{f\theta} &= mr\ddot{\theta} + 2m\dot{r}\dot{\theta} ,
\end{aligned}$$

one arrives at 2 ordinary differential equations for r and θ. The solution of these equations determines the motion of the particle relative to the surface.

4.7 Summary

Two types of forces were discussed in this chapter: friction forces and spring forces. For the friction forces, two classes need to be considered; namely, static friction forces and dynamic friction forces. In addition, it was necessary to consider the two cases of a particle moving relative to a curve and a particle moving relative to a surface.

For a particle moving relative to a curve, triads, $\{\mathbf{e}_t, \mathbf{n}_1, \mathbf{n}_2\}$, associated with each point of the curve are defined. It should be noted that the vectors $\mathbf{e}_n$ and $\mathbf{e}_b$ will lie in the plane spanned by $\mathbf{n}_1$ and $\mathbf{n}_2$. The velocity vector of the point of the curve which is in contact with the particle was denoted $\mathbf{v}_c$. Hence, the relative velocity vector of the particle relative to its point of contact with the curve is $\mathbf{v}_{\mathrm{rel}} = \mathbf{v} - \mathbf{v}_c = v_t \mathbf{e}_t$. The friction force in this case has one component, $\mathbf{F}_f = F_f \mathbf{e}_t$, while the normal force has two components $\mathbf{N} = N_1 \mathbf{n}_1 + N_2 \mathbf{n}_2$.

For a particle moving relative to a surface, triads, $\{\mathbf{n}, \mathbf{t}_1, \mathbf{t}_2\}$, associated with each point of the surface are defined. The velocity vector of the point of the surface which is in contact with the particle was denoted $\mathbf{v}_s$. Hence, the relative velocity vector of the particle relative to its point of contact with the surface is $\mathbf{v}_{\text{rel}} = \mathbf{v} - \mathbf{v}_s = v_1 \mathbf{t}_1 + v_2 \mathbf{t}_2$. The friction force in this case has two components, $\mathbf{F}_f = F_{f1} \mathbf{t}_1 + F_{f2} \mathbf{t}_2$, while the normal force has one component $\mathbf{N} = N\mathbf{n}$.

For both cases, when $\mathbf{v}_{\text{rel}}$ is non-zero, the friction force is of the dynamic Coulomb friction type:

$$\mathbf{F}_f = -\mu_d \|\mathbf{N}\| \frac{\mathbf{v}_{\text{rel}}}{\|\mathbf{v}_{\text{rel}}\|} \, .$$

However, if there is no relative motion, then $\mathbf{v}_{\text{rel}} = \mathbf{0}$, and $\mathbf{F}_f$ is indeterminate. In other words, it must be calculated from $\mathbf{F} = m\mathbf{a}$. However, the magnitude of the friction force is limited by the static friction criterion $\|\mathbf{F}_f\| \le \mu_s \|\mathbf{N}\|$.

The second force we examined was the spring force. We considered a linear spring of stiffness K and unstretched length L. One end of the spring was attached to a fixed point A, while the other end was attached to a particle. Using Hooke's law, the spring force was shown to be

$$\mathbf{F}_s = -K \left(\|\mathbf{r} - \mathbf{r}_A\| - L \right) \frac{\mathbf{r} - \mathbf{r}_A}{\|\mathbf{r} - \mathbf{r}_A\|} \, .$$

Finally, several examples were presented to illustrate the use of the aforementioned expressions for the spring and friction forces.

4.8   Exercises

The following short exercises are intended to assist you in reviewing Chapter 4.

4.1 Consider a particle moving on a circle of radius R. The position vector of the particle is $\mathbf{r} = R\mathbf{e}_r$. Show that the dynamic friction force and normal force have the following representations:

$$\mathbf{F}_f = -\mu_d \|\mathbf{N}\| \frac{\dot{\theta}}{|\dot{\theta}|} \mathbf{e}_\theta , \quad \mathbf{N} = N_r \mathbf{e}_r + N_z \mathbf{E}_z \, .$$

4.2 Consider a particle moving on a cylinder of radius R. The position vector of the particle is $\mathbf{r} = R\mathbf{e}_r + z\mathbf{E}_z$. Show that the dynamic friction force and normal force have the following representations:

$$\mathbf{F}_f = -\mu_d \|\mathbf{N}\| \frac{R\dot{\theta}\mathbf{e}_\theta + \dot{z}\mathbf{E}_z}{\sqrt{R^2\dot{\theta}^2 + \dot{z}^2}} \mathbf{e}_\theta , \quad \mathbf{N} = N\mathbf{e}_r \, .$$

4.3 Consider a particle which is stationary on a rough circle of radius R. The position vector of the particle is $\mathbf{r} = R\mathbf{e}_r$. Show that the static friction force and normal force have the following representations:

$$\mathbf{F}_f = F_f\mathbf{e}_\theta, \quad \mathbf{N} = N_r\mathbf{e}_r + N_z\mathbf{E}_z.$$

In addition, show that the static friction criterion for this problem is

$$|F_f| \leq \mu_s\sqrt{N_r^2 + N_z^2}.$$

Show that this inequality is equivalent to

$$-\mu_s\sqrt{N_r^2 + N_z^2} \leq F_f \leq \mu_s\sqrt{N_r^2 + N_z^2}.$$

4.4 Consider a particle which is stationary on a rough cylinder of radius R. The position vector of the particle is $\mathbf{r} = R\mathbf{e}_r + z\mathbf{E}_z$. Show that the static friction force and normal force have the following representations:

$$\mathbf{F}_f = F_{f\theta}\mathbf{e}_\theta + F_{fz}\mathbf{E}_z, \quad \mathbf{N} = N\mathbf{e}_r.$$

In addition, show that the static friction criterion for this problem is

$$\sqrt{F_{f\theta}^2 + F_{fz}^2} \leq \mu_s|N|.$$

4.5 A particle of mass m is connected to a fixed point O by a spring of stiffness K and unstretched length L. The particle is free to move on a circle of radius R which lies on the $z = z_0$ plane. A vertical gravitational force $-mg\mathbf{E}_z$ acts on the particle. If the circle is smooth, show that the resultant force $\mathbf{F}$ acting on the particle has the representation

$$\mathbf{F} = -mg\mathbf{E}_z + N_z\mathbf{E}_z + N_r\mathbf{e}_r - K\left(\sqrt{R^2 + z_0^2} - L\right)\frac{R\mathbf{e}_r + z_0\mathbf{E}_z}{\sqrt{R^2 + z_0^2}}.$$

4.6 Suppose that the particle in Exercise 4.5 is moving on a rough circle. Show that the resultant force $\mathbf{F}$ acting on the particle now has the representation

$$\begin{aligned}\mathbf{F} &= -mg\mathbf{E}_z + N_z\mathbf{E}_z + N_r\mathbf{e}_r - K\left(\sqrt{R^2 + z_0^2} - L\right)\frac{R\mathbf{e}_r + z_0\mathbf{E}_z}{\sqrt{R^2 + z_0^2}}\\ &\quad - \mu_d\sqrt{N_r^2 + N_z^2}\frac{\dot\theta}{|\dot\theta|}\mathbf{e}_\theta.\end{aligned}$$

4.7 For the dynamic friction force it is a common error to write $\mathbf{F}_f = -\mu_d\|\mathbf{N}\|\frac{\mathbf{v}}{\|\mathbf{v}\|}$. Give two examples which illustrate that this expression is incorrect when the surface or curve that the particle is moving on is itself in motion.

4.8 A particle of mass m is connected to a fixed point O by a spring of stiffness K and unstretched length L. Show that the spring force has the representations

$$\mathbf{F}_s = -K\left(\sqrt{x^2+y^2+z^2} - L\right)\frac{x\mathbf{E}_x + y\mathbf{E}_y + z\mathbf{E}_z}{\sqrt{x^2+y^2+z^2}}$$

$$= -K\left(\sqrt{r^2+z^2} - L\right)\frac{r\mathbf{e}_r + z\mathbf{E}_z}{\sqrt{r^2+z^2}}.$$

4.9 For the system considered in Section 6, establish the differential equations governing the motion of the particle when the contact between the cone and the particle is smooth. In addition, specialize your results to the case where the spring is replaced by an inextensible cable of length L.

5

Power, Work, and Energy

TOPICS

We begin here by defining the mechanical power of a force, and from this the work done by the force during the motion of a particle. Next, the work-energy theorem $\dot{T} = \mathbf{F} \cdot \mathbf{v}$ is derived from the balance of linear momentum. It is then appropriate to discuss conservative forces, and we spend some added time discussing the potential energies of gravitational and spring forces. With these preliminaries aside, energy conservation is discussed. Finally, some examples are presented that show how all of these ideas are used.

5.1   The Power of a Force

Consider a force $\mathbf{P}$ acting on a particle of mass m. The particle has a position vector $\mathbf{r}$ and an absolute velocity vector $\mathbf{v}$. The rate of work done by the force $\mathbf{P}$ on the particle is known as its mechanical power:

$$\text{Mechanical Power of } \mathbf{P} = \mathbf{P} \cdot \mathbf{v}.$$

Consequently, if $\mathbf{P} \cdot \mathbf{v} = 0$, then the force $\mathbf{P}$ does no work.

Consider the work done by $\mathbf{P}$ as the particle moves from $\mathbf{r} = \mathbf{r}(t_A) = \mathbf{r}_A$ to $\mathbf{r} = \mathbf{r}(t_B) = \mathbf{r}_B$ (cf. Figure 5.1). At A, the arc-length parameter $s = s_A$, while at B, $s = s_B$. In what follows, the vector $\mathbf{e}_t$ is the unit tangent vector to the path of the particle. If the particle is moving on a fixed curve, then this vector is also the unit tangent vector to the fixed curve. On the other

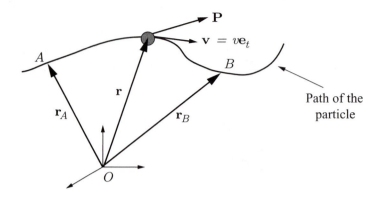

FIGURE 5.1. Schematic of the motion of a particle

hand, if the particle is in motion on a moving curve, then the respective tangent vectors to the curve and the path of the particle will not coincide.[1]

With some algebra, we obtain several equivalent expressions for the work done by $\mathbf{P}$ by integrating its power with respect to time:

$$
\begin{aligned}
W_{AB} &= \int_{t_A}^{t_B} \mathbf{P} \cdot \frac{d\mathbf{r}}{dt} dt \\
&= \int_{\mathbf{r}_A}^{\mathbf{r}_B} \mathbf{P} \cdot d\mathbf{r} \\
&= \int_{t_A}^{t_B} \mathbf{P} \cdot \frac{ds}{dt} \mathbf{e}_t dt \\
&= \int_{s_A}^{s_B} \mathbf{P} \cdot \mathbf{e}_t ds \, .
\end{aligned}
$$

We see from these results, that only the tangential component of $\mathbf{P}$ does work. In particular, forces which are normal to the path of the particle do no work.

Writing the force $\mathbf{P}$ and the differential $d\mathbf{r} = \mathbf{v}dt$ with respect to their components in various bases,

$$
\begin{aligned}
\mathbf{P} &= P_x \mathbf{E}_x + P_y \mathbf{E}_y + P_z \mathbf{E}_z \\
&= P_r \mathbf{e}_r + P_\theta \mathbf{e}_\theta + P_z \mathbf{E}_z \\
&= P_t \mathbf{e}_t + P_n \mathbf{e}_n + P_b \mathbf{e}_b \, , \\
d\mathbf{r} &= dx \mathbf{E}_x + dy \mathbf{E}_y + dz \mathbf{E}_z \\
&= dr \mathbf{e}_r + r d\theta \mathbf{e}_\theta + dz \mathbf{E}_z = ds \mathbf{e}_t \, ,
\end{aligned}
$$

---

[1]The easiest example that illuminates this point is to consider a particle moving on a horizontal line. The tangent vector is constant, say $\mathbf{E}_x$. However, if the line is moving, say with a velocity $v_c \mathbf{E}_y$, then the velocity vector of the particle is not $v \mathbf{E}_x$, rather, it is $v_x \mathbf{E}_x + v_c \mathbf{E}_y$.

we obtain the component forms of the previous results:

$$W_{AB} = \int_{\mathbf{r}_A}^{\mathbf{r}_B} \mathbf{P} \cdot d\mathbf{r} = \int_{\mathbf{r}_A}^{\mathbf{r}_B} P_x dx + P_y dy + P_z dz$$

$$= \int_{\mathbf{r}_A}^{\mathbf{r}_B} P_r dr + P_\theta r d\theta + P_z dz$$

$$= \int_{s_A}^{s_B} \mathbf{P} \cdot \mathbf{e}_t ds$$

$$= \int_{s_A}^{s_B} P_t ds \,.$$

These integrals are evaluated along the path of the particle.

5.2 The Work-Energy Theorem

The kinetic energy T of a particle is defined to be

$$T = \frac{1}{2}m\mathbf{v} \cdot \mathbf{v} = \frac{1}{2}mv^2 = \frac{1}{2}m\left(v_x^2 + v_y^2 + v_z^2\right) \,,$$

where $\mathbf{v} = v\mathbf{e}_t$. The work-energy theorem relates the change in kinetic energy to the resultant force $\mathbf{F}$ acting on the particle:

$$\frac{dT}{dt} = \frac{d}{dt}\left(\frac{1}{2}m\mathbf{v} \cdot \mathbf{v}\right) = \frac{1}{2}m\dot{\mathbf{v}} \cdot \mathbf{v} + \frac{1}{2}m\mathbf{v} \cdot \dot{\mathbf{v}} = m\dot{\mathbf{v}} \cdot \mathbf{v} = \mathbf{F} \cdot \mathbf{v} \,.$$

In sum, the work-energy theorem is

$$\frac{dT}{dt} = \mathbf{F} \cdot \mathbf{v} \,.$$

In words, the rate of change of the kinetic energy is equal to the mechanical power of the resultant force. You should note that this theorem is a consequence of $\mathbf{F} = m\mathbf{a}$.

We can integrate $\dot{T} = \mathbf{F} \cdot \mathbf{v}$ with respect to time to get a result that you have used before:

$$\int_{t_A}^{t_B} \dot{T} dt = \frac{1}{2}mv_B^2 - \frac{1}{2}mv_A^2$$

$$= \int_{t_A}^{t_B} \mathbf{F} \cdot \mathbf{v} dt = \int_{s_A}^{s_B} \mathbf{F} \cdot \mathbf{e}_t ds$$

$$= \int_{s_A}^{s_B} m\mathbf{a} \cdot \mathbf{e}_t ds$$

$$= m \int_{s_A}^{s_B} a_t ds \,.$$

That is,

$$v_B^2 - v_A^2 = \int_{s_A}^{s_B} 2a_t ds\,.$$

Alternatively, you could derive this result from $a_t = \dot{v} = v\frac{dv}{ds}$. Logically, we have just worked backwards through the proof of the work-energy theorem, by substituting $m\mathbf{a}$ for $\mathbf{F}$.

5.3   Conservative Forces

A force $\mathbf{P}$ is defined to be conservative when

$$\mathbf{P} = -\mathrm{grad}(U) = -\frac{\partial U}{\partial \mathbf{r}}\,.$$

Here, $U = U(\mathbf{r})$ is the potential energy of the force $\mathbf{P}$, and the negative sign is a historical convention.

The potential energy function U has several representations:

$$U = U(\mathbf{r}) = \bar{U}(s) = \hat{U}(x, y, z) = \tilde{U}(r, \theta, z)\,.$$

Further, this energy is defined modulo an arbitrary additive constant. Here, we always take this constant to be zero. The gradient of U has the following representations in different coordinate systems:

$$
\begin{aligned}
\frac{\partial U}{\partial \mathbf{r}} &= \frac{\partial \hat{U}}{\partial x}\mathbf{E}_x + \frac{\partial \hat{U}}{\partial y}\mathbf{E}_y + \frac{\partial \hat{U}}{\partial z}\mathbf{E}_z \\
&= \frac{\partial \tilde{U}}{\partial r}\mathbf{e}_r + \frac{1}{r}\frac{\partial \tilde{U}}{\partial \theta}\mathbf{e}_\theta + \frac{\partial \tilde{U}}{\partial z}\mathbf{E}_z \\
&= \frac{\partial \bar{U}}{\partial s}\mathbf{e}_t\,.
\end{aligned}
$$

Further, you can use any of these representations to obtain the results discussed below. However, it is easier to derive most of them without specifying a particular basis or coordinate system.

Physically, if $\mathbf{P}$ is conservative, then the work done by $\mathbf{P}$ in any motion depends only on the endpoints and not on the path. To see this, we use our earlier results on the work done by $\mathbf{P}$:

$$
\begin{aligned}
W_{AB} &= \int_{t_A}^{t_B} \dot{W}\,dt = \int_{t_A}^{t_B} \mathbf{P}\cdot\frac{d\mathbf{r}}{dt}\,dt \\
&= \int_{\mathbf{r}_A}^{\mathbf{r}_B} \mathbf{P}\cdot d\mathbf{r} = -\int_{\mathbf{r}_A}^{\mathbf{r}_B} \frac{\partial U}{\partial \mathbf{r}}\cdot d\mathbf{r} \\
&= U(\mathbf{r}_A) - U(\mathbf{r}_B)\,.
\end{aligned}
$$

Hence, if A and B have the same position vector, then no work is done by **P**. This leads to the statement that the work done by a conservative force in a closed path of the particle is zero.

It is important to note that if **P** is conservative, then its mechanical power has a simple expression:

$$\mathbf{P} \cdot \mathbf{v} = -\frac{\partial U}{\partial \mathbf{r}} \cdot \mathbf{v} = -\frac{dU}{dt}.$$

Not all forces are conservative. For example, tension forces in inextensible strings, friction forces $\mathbf{F}_f$, and normal forces **N** are not conservative.

5.4   Examples of Conservative Forces

The two main examples of conservative forces one encounters are constant forces **C**, of which the gravitational force is an example, and spring forces $\mathbf{F}_s$.

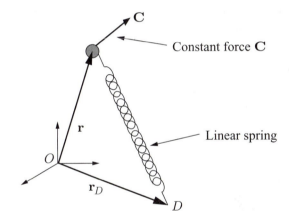

FIGURE 5.2. A particle under the influence of conservative forces **C** and $\mathbf{F}_s$

5.4.1   Constant Forces

The conservative nature of the gravitational force $-mg\mathbf{E}_z$ can be seen by defining the potential energy U_c of any constant force **C**:

$$U_c = -\mathbf{C} \cdot \mathbf{r}.$$

To see that U_c is indeed the potential energy of the force **C**, we need to show that, for all velocity vectors **v**,

$$\mathbf{C} \cdot \mathbf{v} = -\frac{\partial U_c}{\partial \mathbf{r}} \cdot \mathbf{v} = -\frac{dU_c}{dt}.$$

Let's do this:

$$\frac{dU_c}{dt} = \frac{d}{dt}(-\mathbf{C} \cdot \mathbf{r}) = -\dot{\mathbf{C}} \cdot \mathbf{r} - \mathbf{C} \cdot \mathbf{v} = -\mathbf{C} \cdot \mathbf{v}.$$

Based on this result, we then have the following representative constant forces and their potential energies:

$$\mathbf{C} = -mg\mathbf{E}_y \quad \text{and} \quad U_c = mg\mathbf{E}_y \cdot \mathbf{r},$$
$$\mathbf{C} = -mg\mathbf{E}_z \quad \text{and} \quad U_c = mg\mathbf{E}_z \cdot \mathbf{r},$$
$$\mathbf{C} = 10\mathbf{E}_x \quad \text{and} \quad U_c = -10\mathbf{E}_x \cdot \mathbf{r}.$$

You can easily construct some others. Notice that the gravitational potential energy is "mg" times the "height."

5.4.2 Spring Forces

Consider a linear spring of stiffness K and unstretched length L. As shown in Figure 5.2, one end of the spring is attached to a particle of mass m, while the other end is attached to a fixed point D. As shown in the previous chapter, the force exerted by the spring on the particle is

$$\mathbf{F}_s = -K\left(\|\mathbf{r} - \mathbf{r}_D\| - L\right) \frac{\mathbf{r} - \mathbf{r}_D}{\|\mathbf{r} - \mathbf{r}_D\|}.$$

This force is conservative and has a potential energy

$$U_s = \frac{K}{2}\left(\|\mathbf{r} - \mathbf{r}_D\| - L\right)^2.$$

That is the potential energy of a linear spring is half the stiffness times the change in length squared.

To show that U_s is indeed the potential energy of $\mathbf{F}_s$, we need to determine its gradient or, equivalently, we need to show that, for all velocity vectors $\mathbf{v}$,

$$\mathbf{F}_s \cdot \mathbf{v} = -\frac{\partial U_s}{\partial \mathbf{r}} \cdot \mathbf{v} = -\frac{dU_s}{dt}.$$

Because $\mathbf{F}_s$ is not constant,

$$\mathbf{F}_s \cdot \mathbf{v} \neq \frac{d}{dt}(\mathbf{F}_s \cdot \mathbf{r}),$$

and, as a result, this is a difficult result to establish.

To show that U_s is the correct potential energy function, we first need an intermediate result. Suppose $\mathbf{x}$ is a function of time, then

$$\frac{d\|\mathbf{x}\|}{dt} = \frac{d}{dt}\left(\sqrt{\mathbf{x} \cdot \mathbf{x}}\right)$$

$$= \frac{1}{2\sqrt{\mathbf{x} \cdot \mathbf{x}}} \frac{d}{dt} (\mathbf{x} \cdot \mathbf{x})$$

$$= \frac{1}{2\sqrt{\mathbf{x} \cdot \mathbf{x}}} (\dot{\mathbf{x}} \cdot \mathbf{x} + \mathbf{x} \cdot \dot{\mathbf{x}})$$

$$= \frac{\dot{\mathbf{x}} \cdot \mathbf{x}}{\sqrt{\mathbf{x} \cdot \mathbf{x}}}$$

$$= \frac{\dot{\mathbf{x}} \cdot \mathbf{x}}{\|\mathbf{x}\|}.$$

If we let $\mathbf{x} = \mathbf{r} - \mathbf{r}_D$, then we find the result

$$\frac{d\|\mathbf{r} - \mathbf{r}_D\|}{dt} = \frac{(\dot{\mathbf{r}} - \dot{\mathbf{r}}_D) \cdot (\mathbf{r} - \mathbf{r}_D)}{\|\mathbf{r} - \mathbf{r}_D\|} = \frac{\mathbf{v} \cdot (\mathbf{r} - \mathbf{r}_D)}{\|\mathbf{r} - \mathbf{r}_D\|}.$$

Let's now differentiate U_s:

$$\frac{dU_s}{dt} = \frac{d}{dt} \left(\frac{K}{2} (\|\mathbf{r} - \mathbf{r}_D\| - L)^2 \right)$$

$$= K (\|\mathbf{r} - \mathbf{r}_D\| - L) \frac{d}{dt} (\|\mathbf{r} - \mathbf{r}_D\| - L)$$

$$= K (\|\mathbf{r} - \mathbf{r}_D\| - L) \frac{\dot{\mathbf{r}} \cdot (\mathbf{r} - \mathbf{r}_D)}{\|\mathbf{r} - \mathbf{r}_D\|}$$

$$= -\mathbf{F}_s \cdot \mathbf{v}.$$

Hence, U_s is indeed the potential energy of the linear spring force. You should notice that in the proof we used the fact that $\mathbf{v}_D = \mathbf{0}$. Finally, the expression we have obtained for U_s is valid even when $\mathbf{r}_D = \mathbf{0}$, that is, when the point D can be chosen to be the origin.

5.5   Energy Conservation

Consider a particle of mass m that is acted upon by a set of forces: n of these forces are conservative, $\mathbf{F}_1, \mathbf{F}_2, \ldots, \mathbf{F}_n$, and the remainder, whose resultant we denote by $\mathbf{F}_{nc}$, are nonconservative. The potential energies of the conservative forces are denoted by $U_1, U_2, \ldots, U_n$:

$$\mathbf{F}_i = -\frac{\partial U_i}{\partial \mathbf{r}},$$

where $i = 1, \ldots, n$. The resultant conservative force acting on the particle is

$$\mathbf{F}_c = \sum_{i=1}^{n} \mathbf{F}_i = -\sum_{i=1}^{n} \frac{\partial U_i}{\partial \mathbf{r}} = -\frac{\partial U}{\partial \mathbf{r}},$$

where U, the total potential energy of the conservative forces, is

$$U = \sum_{i=1}^{n} U_i .$$

In summary,

$$\mathbf{F} = \mathbf{F}_{nc} + \mathbf{F}_c = \mathbf{F}_{nc} - \frac{\partial U}{\partial \mathbf{r}} .$$

To establish energy-conservation results, we start with the work-energy theorem:

$$\frac{dT}{dt} = \mathbf{F} \cdot \mathbf{v} .$$

For the case at hand,

$$
\begin{aligned}
\frac{dT}{dt} &= \mathbf{F} \cdot \mathbf{v} = (\mathbf{F}_c + \mathbf{F}_{nc}) \cdot \mathbf{v} \\
&= \left(-\frac{\partial U}{\partial \mathbf{r}} + \mathbf{F}_{nc} \right) \cdot \mathbf{v} \\
&= -\frac{dU}{dt} + \mathbf{F}_{nc} \cdot \mathbf{v} .
\end{aligned}
$$

We now define the total energy E of the particle:

$$E = T + U .$$

This energy is the sum of the kinetic and potential energies of the particle. Rearranging the previous equation with the assistance of the definition of E, we find that

$$\frac{dE}{dt} = \mathbf{F}_{nc} \cdot \mathbf{v} .$$

This equation can be viewed as an alternative form of the work-energy theorem.

If the nonconservative forces do no work during a motion of the particle, that is, $\mathbf{F}_{nc} \cdot \mathbf{v} = \mathbf{0}$, then the total energy E of the particle is conserved:[2]

$$\frac{dE}{dt} = 0 .$$

This implies that E is a constant E_0 during the motion of the particle:[3]

$$E = T + U = \frac{1}{2}mv^2 + U(\mathbf{r}) = E_0 .$$

---

[2]This is a classical result that was known, although not in the form written here, to the Dutch scientist Christiaan Huygens (1629–1695) and the German scientist Gottfried Wilhelm Leibniz (1646–1716). These men were contemporaries of Isaac Newton (1643–1727).

[3]A common error is to assume that $\dot{E} = 0$ implies that $E = 0$. It does not.

During an energy-conserving motion, there is a transfer between the kinetic
and potential energies of a particle.

In problems, one uses energy conservation $\dot{E} = 0$ to solve for one unknown. For example, suppose one is given an initial speed v_0 and location
$\mathbf{r}_0$ of a particle at some instant during an energy-conserving motion. One
can use energy conservation to determine the speed v at another location
$\mathbf{r}$ during this motion:

$$v^2 = \frac{2}{m} \left(U(\mathbf{r}_0) - U(\mathbf{r}) \right) + v_0^2 \,.$$

In light of our earlier discussion in Section 2 of the identity $a_t = \frac{dv}{dt} = v\frac{dv}{ds}$,
you should notice that

$$\frac{1}{m} \left(U(\mathbf{r}_0) - U(\mathbf{r}) \right) = \int_{s_0}^{s} a_t(u)du = \frac{1}{m} \int_{t_0}^{t} \mathbf{F} \cdot \mathbf{v} d\tau \,.$$

That is, if the only forces that do work are conservative, then the existence
of the potential energies makes integrating the power of $\mathbf{F}$ trivial.

5.6   A Particle Moving on a Rough Curve

Consider a particle of mass m that is moving on the rough space curve
shown in Figure 5.3. A gravitational force $-mg\mathbf{E}_z$ acts on the particle. In
addition, a linear spring of stiffness K and unstretched length L is attached
to the particle and a fixed point C.

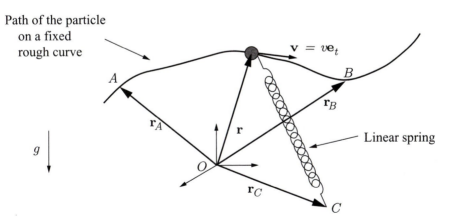

FIGURE 5.3. A particle in motion on a fixed space curve

We wish to determine the work done by the friction force on the particle
as it moves from $\mathbf{r} = \mathbf{r}(t_A) = \mathbf{r}_A$ to $\mathbf{r} = \mathbf{r}(t_B) = \mathbf{r}_B$. Furthermore, we ask
the question, if the curve were smooth, then given $\mathbf{v}(t_A)$, what is $\mathbf{v}(t_B)$?

5.6.1   Kinematics

Here, we will use the Serret-Frenet triads, and we tacitly assume that these vectors can be calculated for the curve at hand. We then have the usual results

$$\mathbf{v} = v\mathbf{e}_t \,, \quad \mathbf{a} = \dot{v}\mathbf{e}_t + \kappa v^2 \mathbf{e}_n \,, \quad T = \frac{1}{2}mv^2 \,.$$

5.6.2   Forces

We leave the free-body diagram as an exercise, and record that the resultant force $\mathbf{F}$ acting on the particle is

$$\mathbf{F} = \mathbf{F}_f + N_n\mathbf{e}_n + N_b\mathbf{e}_b - mg\mathbf{E}_z - K\left(\|\mathbf{r} - \mathbf{r}_C\| - L\right)\frac{\mathbf{r} - \mathbf{r}_C}{\|\mathbf{r} - \mathbf{r}_C\|} \,.$$

If the friction is dynamic Coulomb friction, then

$$\mathbf{F}_f = -\mu_d\|N_n\mathbf{e}_n + N_b\mathbf{e}_b\|\frac{\mathbf{v}}{\|\mathbf{v}\|} \,.$$

On the other hand, if the particle is stationary, then the friction is static.

5.6.3   Work Done by Friction

We are now in a position to start from the work-energy theorem and establish how the rate of change of total energy is related to the power of the friction force:

$$\begin{aligned}
\frac{dT}{dt} &= \mathbf{F} \cdot \mathbf{v} \\
&= \mathbf{F}_f \cdot \mathbf{v} + N_n\mathbf{e}_n \cdot \mathbf{v} + N_b\mathbf{e}_b \cdot \mathbf{v} - mg\mathbf{E}_z \cdot \mathbf{v} \\
&\quad - K\left(\|\mathbf{r} - \mathbf{r}_C\| - L\right)\frac{\mathbf{r} - \mathbf{r}_C}{\|\mathbf{r} - \mathbf{r}_C\|} \cdot \mathbf{v} \,.
\end{aligned}$$

However, the normal forces are perpendicular to the velocity vector, and the spring and gravitational forces are conservative:

$$\begin{aligned}
N_n\mathbf{e}_n \cdot \mathbf{v} &= 0 \,, \\
N_b\mathbf{e}_b \cdot \mathbf{v} &= 0 \,, \\
-mg\mathbf{E}_z \cdot \mathbf{v} &= -\frac{d}{dt}\left(mg\mathbf{E}_z \cdot \mathbf{r}\right) \,, \\
-K\left(\|\mathbf{r} - \mathbf{r}_C\| - L\right)\frac{\mathbf{r} - \mathbf{r}_C}{\|\mathbf{r} - \mathbf{r}_C\|} \cdot \mathbf{v} &= -\frac{d}{dt}\left(\frac{K}{2}\left(\|\mathbf{r} - \mathbf{r}_C\| - L\right)^2\right) \,.
\end{aligned}$$

As a result,

$$\frac{dE}{dt} = \mathbf{F}_f \cdot \mathbf{v} \,,$$

where the total energy of the particle E is

$$E = T + \frac{K}{2} (\|\mathbf{r} - \mathbf{r}_C\| - L)^2 + mg\mathbf{E}_z \cdot \mathbf{r}.$$

You could also have arrived at this result by starting with the alternative form of the work-energy theorem, $\dot{E} = \mathbf{F}_{nc} \cdot \mathbf{v}$, and noting that the normal forces are perpendicular to the velocity of the particle and hence are workless.

The work done by the friction force can be found by integrating the above equation for the time rate of change of the total energy E:

$$\begin{aligned}
\int_{t_A}^{t_B} \mathbf{F}_f \cdot \mathbf{v}dt &= E_B - E_A \\
&= \frac{1}{2}m \left(v_B^2 - v_A^2\right) + mg\mathbf{E}_z \cdot (\mathbf{r}_B - \mathbf{r}_A) \\
&\quad + \frac{K}{2} \left((\|\mathbf{r}_B - \mathbf{r}_C\| - L)^2 - (\|\mathbf{r}_A - \mathbf{r}_C\| - L)^2\right).
\end{aligned}$$

Clearly, if you know $\mathbf{r}_A$, $\mathbf{r}_B$, v_A, and v_B, then you don't need to directly integrate $\mathbf{F}_f \cdot \mathbf{v}$ to determine the work done by the friction force. Furthermore, given the above information, one doesn't need to calculate the Serret-Frenet triads for each point along the curve.

It is interesting to note that if the friction were of the dynamic Coulomb type, then

$$\begin{aligned}
\frac{dE}{dt} &= \mathbf{F}_f \cdot \mathbf{v} = -\mu_d \|N_n \mathbf{e}_n + N_b \mathbf{e}_b\| \frac{\mathbf{v}}{\|\mathbf{v}\|} \cdot \mathbf{v} \\
&= -\mu_d \|N_n \mathbf{e}_n + N_b \mathbf{e}_b\| \|\mathbf{v}\| < 0.
\end{aligned}$$

In other words, such a friction force will dissipate energy, as expected.

5.6.4   The Smooth Curve

For a smooth curve, $\mathbf{F}_f = \mathbf{0}$, and we have energy conservation $\dot{E} = 0$. In this case, given $\mathbf{r}_A$, $\mathbf{r}_B$, and v_A, one can calculate v_B:

$$v_B^2 = v_A^2 + 2g\mathbf{E}_z \cdot (\mathbf{r}_A - \mathbf{r}_B) + \frac{K}{2} \left((\|\mathbf{r}_A - \mathbf{r}_C\| - L)^2 - (\|\mathbf{r}_B - \mathbf{r}_C\| - L)^2\right).$$

This is the main use of energy conservation in the problems discussed in engineering dynamics courses. Notice that because the only forces that do work on the particle are conservative, the velocity at B does not depend on the path between A and B.

A specific example of this case was discussed in Section 5 of Chapter 3. In this example, the spring force was absent, and it was shown there, without using the work-energy theorem directly, that

$$v^2(x) = v^2(x_0) + \int_{x_0}^{x} 4ugdu = v^2(x_0) + 2g\left(x^2 - x_0^2\right).$$

Our earlier comment that this was an energy conservation result should now be obvious to you.

5.7   Further Examples of Energy Conservation

We can use the results of the previous problem for many others simply by doing some minor modifications.

5.7.1   A Particle in a Conservative Force Field

Consider a particle moving in space, as opposed to along a fixed curve. The particle is assumed to be under the sole influences of a gravitational force and a spring force:

$$\mathbf{F} = -mg\mathbf{E}_z - K\left(\|\mathbf{r} - \mathbf{r}_C\| - L\right) \frac{\mathbf{r} - \mathbf{r}_C}{\|\mathbf{r} - \mathbf{r}_C\|}.$$

By setting $\mathbf{F}_f = \mathbf{0}$ and $\mathbf{N} = \mathbf{0}$ in the previous example, one finds that the total energy E of the particle is conserved, where

$$E = T + \frac{K}{2}\left(\|\mathbf{r} - \mathbf{r}_C\| - L\right)^2 + mg\mathbf{E}_z \cdot \mathbf{r}.$$

The Serret-Frenet triad for this example pertains to the path of the particle as opposed to the prescribed fixed curve, and it is impossible to explicitly determine this triad without first solving $\mathbf{F} = m\mathbf{a}$ for the motion $\mathbf{r}(t)$ of the particle.

5.7.2   A Particle on a Fixed Smooth Surface

If a particle is moving on any fixed smooth surface under conservative gravitational and spring forces, then

$$\mathbf{F} = -mg\mathbf{E}_z - K\left(\|\mathbf{r} - \mathbf{r}_C\| - L\right) \frac{\mathbf{r} - \mathbf{r}_C}{\|\mathbf{r} - \mathbf{r}_C\|} + \mathbf{N}.$$

Here $\mathbf{N}$ is the normal (or reaction) force exerted by the surface on the particle. However, since this force is perpendicular to the velocity vector of the particle, $\mathbf{N} \cdot \mathbf{v} = 0$. Consequently, the total energy

$$E = T + \frac{K}{2}\left(\|\mathbf{r} - \mathbf{r}_C\| - L\right)^2 + mg\mathbf{E}_z \cdot \mathbf{r}$$

is again conserved.

5.7.3   The Planar Pendulum

We described the planar pendulum in Section 4 of Chapter 2. For this example, one can show using the work-energy theorem that the total energy E of the particle is conserved. To begin,

$$
\begin{aligned}
\frac{dT}{dt} &= \mathbf{F} \cdot \mathbf{v} \\
&= N_r \mathbf{e}_r \cdot \mathbf{v} - mg\mathbf{E}_y \cdot \mathbf{v} - N\mathbf{E}_z \cdot \mathbf{v} \\
&= -\frac{d}{dt}\left(mg\mathbf{E}_y \cdot \mathbf{r}\right),
\end{aligned}
$$

where we have changed notation and defined $N_r \mathbf{e}_r$ as the tension force in the string/rod. This force and the normal force $N\mathbf{E}_z$ are perpendicular to the velocity vector and, as a result, are workless. It now follows that the total energy E of the particle is conserved:

$$
\frac{d}{dt}\left(E = \frac{1}{2}m\mathbf{v} \cdot \mathbf{v} + mg\mathbf{E}_y \cdot \mathbf{r}\right) = 0.
$$

5.8   Summary

The first concept introduced in this chapter was the mechanical power of a force $\mathbf{P}$ acting on a particle whose absolute velocity vector is $\mathbf{v}$: $\mathbf{P} \cdot \mathbf{v}$. The work done by a force $\mathbf{P}$ in a time-interval $[t_A, t_B]$ is the time-integral of its power:

$$
W_{AB} = \int_{t_A}^{t_B} \mathbf{P} \cdot \mathbf{v} dt.
$$

Dependent on the coordinate system used there are numerous representations of this integral. You should notice that in order to evaluate the integral it is necessary to know the path of the particle. If, for all possible paths, $W_{AB} = 0$, then a force does no work and its power must be zero. In other words, $\mathbf{P}$ must be normal to $\mathbf{v}$.

An important class of forces were then discussed - conservative forces. A force $\mathbf{P}$ is conservative if one can find a potential energy function $U = U(\mathbf{r})$ such that, for all possible motions,

$$
\mathbf{P} = -\frac{\partial U}{\partial \mathbf{r}},
$$

or, equivalently,

$$
\dot{U} = -\mathbf{P} \cdot \mathbf{v}.
$$

Because a conservative force is the gradient of a scalar function $U = U(\mathbf{r})$, the work done by this class of forces is independent of the path of the particle. In Section 4 of this chapter, it was shown that a spring force $\mathbf{F}_s$ and a constant force $\mathbf{C}$ are conservative:

$$\mathbf{F}_s = -K\left(\|\mathbf{r} - \mathbf{r}_D\| - L\right)\frac{\mathbf{r} - \mathbf{r}_D}{\|\mathbf{r} - \mathbf{r}_D\|} = -\frac{\partial U_s}{\partial \mathbf{r}},$$

$$\mathbf{C} = -\frac{\partial U_c}{\partial \mathbf{r}},$$

where

$$U_s = \frac{K}{2}\left(\|\mathbf{r} - \mathbf{r}_D\| - L\right)^2, \quad U_c = -\mathbf{C} \cdot \mathbf{r}.$$

Not all forces are conservative. In particular, friction and normal forces are nonconservative.

Using the notion of mechanical power, two versions of the work-energy theorem were established:

$$\dot{T} = \mathbf{F} \cdot \mathbf{v}, \quad \dot{E} = \mathbf{F}_{nc} \cdot \mathbf{v}.$$

Here, $T = \frac{m}{2}\mathbf{v} \cdot \mathbf{v}$ is the kinetic energy of the particle, $E = T + U$ is the total energy of the particle, U is the sum of the potential energies of all the conservative forces acting on the particle and $\mathbf{F}_{nc}$ is the resultant nonconservative force acting on the particle. You should notice that $\mathbf{F}_{nc} = \mathbf{F} + \frac{\partial U}{\partial \mathbf{r}}$.

The remainder of the chapter was concerned with using the work-energy theorem to examine the work done by friction forces and providing examples of systems where the total energy E was conserved. For all of the examples of energy conservation, the work-energy theorem was used to show that $\mathbf{F}_{nc} \cdot \mathbf{v} = 0$, and, consequently, E must be conserved. As illustrated in the examples of energy conservation, when E is conserved then certain information on the speed of the particle as a function of position can be determined without explicitly integrating the differential equations governing the motion of the particle.

5.9   Exercises

The following short exercises are intended to assist you in reviewing Chapter 5.

5.1 Give examples to illustrate the following statement: "Every constant force is conservative, but not all conservative forces are constant."

5.2 Starting from the definition of the kinetic energy $T = \frac{m}{2}\mathbf{v} \cdot \mathbf{v}$, establish the work-energy theorem $\dot{T} = \mathbf{F} \cdot \mathbf{v}$. Then, using this result, derive the alternative form of the work-energy theorem $\dot{E} = \mathbf{F}_{nc} \cdot \mathbf{v}$.

5.3 Give three examples of particle problems where the total energy E is conserved.

5.4 A particle is moving on a smooth horizontal plane. A gravitational force $-mg\mathbf{E}_z$ acts on the particle. If the plane is given a vertical motion, then why does the normal force acting on the particle perform work? Using this example, show that the normal force is not a conservative force.

5.5 Give three examples of particle problems where the total energy E is not conserved.

5.6 A particle is free to move on a smooth plane $z = 0$. It is attached to a fixed point O by a linear spring of stiffness K and unstretched length L. A gravitational force $-mg\mathbf{E}_z$ acts on the particle. Starting from the work-energy theorem, prove that

$$E = \frac{m}{2}\left(\dot{r}^2 + r^2\dot{\theta}^2\right) + \frac{K}{2}\left(r - L\right)^2$$

is conserved.

5.7 Consider the same system discussed in Exercise 5.6, but in this case assume that the surface is rough. Show that

$$\dot{E} = -\mu_d mg\sqrt{\dot{r}^2 + r^2\dot{\theta}^2}\,.$$

5.8 For any vector $\mathbf{b}$ show that

$$\frac{d}{dt}\|\mathbf{b}\| = \frac{\mathbf{b}\cdot\dot{\mathbf{b}}}{\|\mathbf{b}\|}\,.$$

Using this result, show that if the magnitude of $\mathbf{b}$ is constant, then any change in $\mathbf{b}$ must be normal to $\mathbf{b}$.

5.9 With the assistance of the identity established in Exercise 5.8, show that Newton's gravitational force,

$$\mathbf{F}_N = -\frac{GMm}{\|\mathbf{r}\|^2}\frac{\mathbf{r}}{\|\mathbf{r}\|}\,,$$

is conservative with a potential energy

$$U_N = -\frac{GMm}{\|\mathbf{r}\|}\,.$$

Here, G, M, and m are constants.

6
Momenta, Impulses, and Collisions

TOPICS

Here, the linear and angular momenta of a particle are discussed. In particular, conditions for the conservation of these kinematical quantities are established. This is followed by a discussion of impact problems where particles are used as models for the impacting bodies.

6.1 Linear Momentum and Its Conservation

Consider a particle of mass m moving in space. As usual, the position vector of the particle relative to a fixed origin is denoted by $\mathbf{r}$. We recall that the linear momentum $\mathbf{G}$ of the particle is defined to be

$$\mathbf{G} = m\mathbf{v} = m\dot{\mathbf{r}}.$$

6.1.1 Linear Impulse and Linear Momentum

A more primitive form of the balance of linear momentum $\mathbf{F} = m\mathbf{a}$ is its integral form:

$$\mathbf{G}(t_1) - \mathbf{G}(t_0) = \int_{t_0}^{t_1} \mathbf{F}dt.$$

This equation is assumed to hold for all intervals of time, and hence for all times t_0 and t_1. The time integral of any force is known as its *linear impulse*.

The reason the integral form is more primitive than the equation $\mathbf{F} = m\mathbf{a}$ is that it does not assume that $\mathbf{v}$ can always be differentiated to determine $\mathbf{a}$. We shall use the integral form of the balance of linear momentum in our forthcoming discussion of impact.

6.1.2 Conservation of Linear Momentum

The conservation of some component of linear momentum of the particle is an important feature of many problems. Suppose that the component of $\mathbf{G}$ in a direction of a *given* vector $\mathbf{c}$ is conserved:

$$\frac{d}{dt}\left(\mathbf{G} \cdot \mathbf{c}\right) = 0 .$$

To examine the conditions under which this arises, we expand the time derivative on the right-hand side of the above equation to find that

$$\frac{d}{dt}\left(\mathbf{G} \cdot \mathbf{c}\right) = \dot{\mathbf{G}} \cdot \mathbf{c} + \mathbf{G} \cdot \dot{\mathbf{c}} .$$

Consequently, given a vector $\mathbf{c}$,

$\mathbf{G} \cdot \mathbf{c}$ is conserved if, and only if, $\mathbf{F} \cdot \mathbf{c} + \mathbf{G} \cdot \dot{\mathbf{c}} = 0$.

A special case of this result arises when $\mathbf{c}$ is constant; then the condition for the conservation of the linear momentum in the direction of $\mathbf{c}$ is none other than $\mathbf{F} \cdot \mathbf{c} = 0$. That is, there is no force in this direction.

In general, the most difficult aspect of using the conservation of linear momentum is to find appropriate directions $\mathbf{c}$. This is an art.

6.1.3 Example

You have already seen several examples of linear momentum conservation. For instance, consider a particle moving in a gravitational field. Here, $\mathbf{F} = -mg\mathbf{E}_y$. As a result, $\mathbf{F} \cdot \mathbf{E}_z = 0$ and $\mathbf{F} \cdot \mathbf{E}_x = 0$. So the linear momentum of the particle and, as a result, its velocity in the directions $\mathbf{E}_x$ and $\mathbf{E}_z$ are conserved. It is perhaps instructive to note that the linear momentum of the particle in the $\mathbf{E}_y$ direction is not conserved.

6.2 Angular Momentum and Its Conservation

As shown in Figure 6.1, let $\mathbf{r}$ be the position vector of a particle relative to a fixed point O, and let $\mathbf{v}$ be the absolute velocity vector of the particle. Then the angular momentum $\mathbf{H}_O$ of the particle (relative to O) is

$$\mathbf{H}_O = \mathbf{r} \times m\mathbf{v} = \mathbf{r} \times \mathbf{G} .$$

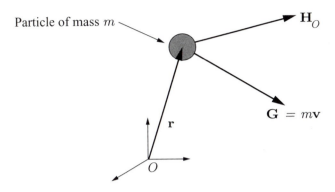

FIGURE 6.1. Some kinematical quantities

In Cartesian coordinates, $\mathbf{H}_O$ has the representation

$$\mathbf{H}_O = \mathbf{r} \times m\mathbf{v} \quad = \quad \det \begin{bmatrix} \mathbf{E}_x & \mathbf{E}_y & \mathbf{E}_z \\ x & y & z \\ m\dot{x} & m\dot{y} & m\dot{z} \end{bmatrix}$$

$$= \quad m(y\dot{z} - z\dot{y})\mathbf{E}_x + m(z\dot{x} - x\dot{z})\mathbf{E}_y + m(x\dot{y} - y\dot{x})\mathbf{E}_z .$$

In cylindrical polar coordinates, $\mathbf{H}_O$ has the representation

$$\mathbf{H}_O = \mathbf{r} \times m\mathbf{v} \quad = \quad \det \begin{bmatrix} \mathbf{e}_r & \mathbf{e}_\theta & \mathbf{E}_z \\ r & 0 & z \\ m\dot{r} & mr\dot{\theta} & m\dot{z} \end{bmatrix}$$

$$= \quad -mzr\dot{\theta}\mathbf{e}_r + m(z\dot{r} - r\dot{z})\mathbf{e}_\theta + mr^2\dot{\theta}\mathbf{E}_z .$$

6.2.1 Angular Momentum Theorem

To determine how the angular momentum changes with time, we invoke the balance of linear momentum:

$$\frac{d\mathbf{H}_O}{dt} = \frac{d}{dt}(\mathbf{r} \times m\mathbf{v}) = \mathbf{v} \times m\mathbf{v} + \mathbf{r} \times m\dot{\mathbf{v}} = \mathbf{r} \times \mathbf{F} .$$

In summary, the angular momentum theorem is

$$\frac{d\mathbf{H}_O}{dt} = \mathbf{r} \times \mathbf{F} .$$

As opposed to our later developments with rigid bodies, a balance of angular momentum $\dot{\mathbf{H}}_O = \mathbf{r} \times \mathbf{F}$ is not an independent postulate. It arises as a consequence of the balance of linear momentum.

6.2.2 Conservation of Angular Momentum

The conservation of some component of angular momentum of the particle is an important feature of many problems. Suppose that the component of

$\mathbf{H}_O$ in the direction of a *given* vector $\mathbf{c}$ is conserved:

$$\frac{d}{dt}(\mathbf{H}_O \cdot \mathbf{c}) = 0 \, .$$

To examine the conditions under which this arises, we expand the time derivative on the left-hand side of the above equation and invoke the angular momentum theorem to find that

$$\begin{aligned} \frac{d}{dt}(\mathbf{H}_O \cdot \mathbf{c}) &= \dot{\mathbf{H}}_O \cdot \mathbf{c} + \mathbf{H}_O \cdot \dot{\mathbf{c}} \\ &= (\mathbf{r} \times \mathbf{F}) \cdot \mathbf{c} + \mathbf{H}_O \cdot \dot{\mathbf{c}} \, . \end{aligned}$$

Consequently, given a vector $\mathbf{c}$,

$\mathbf{H}_O \cdot \mathbf{c}$ is conserved if, and only if, $(\mathbf{r} \times \mathbf{F}) \cdot \mathbf{c} + \mathbf{H}_O \cdot \dot{\mathbf{c}} = 0$.

A special case of this result arises when $\mathbf{c}$ is constant. Then the condition for the conservation of the angular momentum in the direction of $\mathbf{c}$ is none other than $(\mathbf{r} \times \mathbf{F}) \cdot \mathbf{c} = 0$.

In an undergraduate dynamics course, problems where angular momentum is conserved can usually be set up so that $\mathbf{c} = \mathbf{E}_z$. We shall shortly examine such an example.

6.2.3 Central Force Problems

A central force problem is one where $\mathbf{F}$ is parallel to $\mathbf{r}$. From the angular momentum theorem, it follows that $\mathbf{H}_O$ is conserved. This conservation implies several interesting results, which we now discuss.

Since $\mathbf{H}_O$ is conserved, $\mathbf{H}_O = h\mathbf{h}$, where h and $\mathbf{h}$ are constant. We can choose $\mathbf{h}$ to be a unit vector. Because $\mathbf{r} \times m\mathbf{v}$ is constant, the vectors $\mathbf{r}$ and $\mathbf{v}$ form a plane with a constant unit normal vector $\mathbf{h}$. This plane passes through the origin O and is fixed. Given a set of initial conditions $\mathbf{r}(t_0)$ and $\mathbf{v}(t_0)$, we can choose a cylindrical polar coordinate system such that $\mathbf{E}_z = \mathbf{h}$, $\mathbf{r} = r\mathbf{e}_r$, and $\mathbf{v} = \dot{r}\mathbf{e}_r + r\dot{\theta}\mathbf{e}_\theta$. To do this, it suffices to choose $\mathbf{E}_z$ such that the following equation is satisfied:

$$\mathbf{H}_O = h\mathbf{E}_z = \mathbf{r}(t_0) \times \mathbf{v}(t_0) \, .$$

This simplifies the problem of determining the motion of the particle considerably. Furthermore, $h = mr^2\dot{\theta}$ is constant during the motion of the particle.

6.2.4 Kepler's Problem

The most famous example of a central force problem, and angular momentum conservation, was solved by Newton. In seeking to develop a model for planetary motion which would explain Kepler's laws[1] and astronomical observations, he postulated, as a model for the resultant force $\mathbf{F}$ exerted on a planet of mass m by a fixed planet of mass M, the following conservative force:

$$\mathbf{F} = -\frac{GmM}{\|\mathbf{r}\|^2}\frac{\mathbf{r}}{\|\mathbf{r}\|} = -\frac{\partial U}{\partial \mathbf{r}}, \quad U = -\frac{GmM}{\|\mathbf{r}\|}.$$

Here, G is the universal gravitational constant, and the fixed planet is taken to be the origin. Clearly, we can use the previous results on central force problems here.

Using the balance of linear momentum for the planet of mass m, we find two ordinary differential equations:

$$m\ddot{r} - mr\dot{\theta}^2 = -\frac{GMm}{r^2}, \quad mr\ddot{\theta} + 2m\dot{r}\dot{\theta} = 0.$$

The solution of these equations is facilitated by the use of two conserved quantities, the total energy E and the magnitude of the angular momentum h:

$$\begin{aligned} E &= \frac{1}{2}m\mathbf{v}\cdot\mathbf{v} + U = \frac{1}{2}m\left(\dot{r}^2 + r^2\dot{\theta}^2\right) - \frac{GmM}{r}, \\ h &= \mathbf{H}_O \cdot \mathbf{E}_z = mr^2\dot{\theta}. \end{aligned}$$

However, we do not pursue this matter any further here.[2]

6.2.5   A Particle on a Smooth Cone

As shown in Figure 6.2, we return to an example discussed in Section 6 of Chapter 4. Here, however, the surface of the cone is assumed to be smooth. We wish to show that $\mathbf{H}_O \cdot \mathbf{E}_z$ is conserved for the particle.[3]

Let us first recall some kinematical results from Section 6 of Chapter 4:

$$\begin{aligned} \mathbf{r} &= r\mathbf{e}_r + r\tan(\alpha)\mathbf{E}_z, \\ \mathbf{v} &= \dot{r}\left(\mathbf{e}_r + \tan(\alpha)\mathbf{E}_z\right) + r\dot{\theta}\mathbf{e}_\theta. \end{aligned}$$

---

[1]Johannes Kepler (1571–1630) was a German astronomer and physicist who, based on observations of the orbits of certain planets, proposed three laws to explain planetary motion. His three laws are arguable the most famous of his many scientific contributions.

[2]For further treatment of these equations, see, for example, Arnol'd [2] and Moulton [41]. Discussions of this central force problem can also be found in every undergraduate dynamics text, for example, in Section 13, Chapter 3 of Meriam and Kraige [39] and Section 15, Chapter 5 of Riley and Sturges [50]. You should notice that these texts assume that the motion of the particle is planar.

[3]We leave it as an exercise to show that the total energy of the particle is conserved. Another exercise is to show that angular momentum is also conserved if the spring is replaced by an inextensible string.

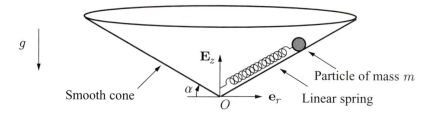

FIGURE 6.2. A particle on a smooth cone

A simple calculation shows that the angular momentum of the particle is

$$\mathbf{H}_O = \mathbf{r} \times m\mathbf{v} = \frac{mr^2\dot{\theta}}{\cos(\alpha)} \left(\cos(\alpha)\mathbf{E}_z - \sin(\alpha)\mathbf{e}_r \right) .$$

We also note that the resultant force acting on the particle is composed of a normal force, gravity, and a spring force:

$$\mathbf{F} = N\mathbf{n} - mg\mathbf{E}_z - K \left(\|\mathbf{r}\| - L \right) \frac{\mathbf{r}}{\|\mathbf{r}\|} .$$

The moment of the resultant force is in the $\mathbf{e}_\theta$ direction:

$$\begin{aligned}
\mathbf{r} \times \mathbf{F} &= \mathbf{r} \times N\mathbf{n} - \mathbf{r} \times mg\mathbf{E}_z - \mathbf{r} \times K \left(\|\mathbf{r}\| - L \right) \frac{\mathbf{r}}{\|\mathbf{r}\|} \\
&= \left(mgr - \frac{Nr}{\cos(\alpha)} \right) \mathbf{e}_\theta .
\end{aligned}$$

Consequently, $\mathbf{H}_0 \cdot \mathbf{E}_z$ is conserved:

$$mr^2\dot{\theta} = \text{constant} .$$

You should notice that during the motion of the particle it is impossible for $\dot{\theta}$ to change sign.

6.3   Collision of Particles

The collision of two bodies involves substantial deformations during the impact and may induce permanent deformations. Ignoring the rotational motion of the bodies, the simplest model to determine the postcollision velocities of the bodies is to use a mass particle to model each individual body. However, mass particles cannot deform, and so one needs to introduce some (seemingly ad hoc) parameter to account for this feature. The parameter commonly used is the coefficient of restitution e.

The theory we present here is often referred to as frictionless, oblique, central impact of two particles. Other theories are available that account for friction and rotational inertias. The reader is referred to Brach [10], Goldsmith [28], Routh [51], Rubin [53], and Stewart [60] for discussions on other theories, applications and unresolved issues. For a discussion of some of Newton's contributions to this subject, see Problem 12 on Pages 148–151 of [44]. There, Newton discusses the impact of two spheres.

6.3.1   The Model and Impact Stages

In what follows, we model two impacting bodies of masses m_1 and m_2 by two mass particles of masses m_1 and m_2, respectively. Further, the bodies are assumed to be in a state of purely translational motion, so the velocity vector of any point of one of the bodies is identical to the velocity of the mass particle modeling the body. The position vector of the mass particle is defined to coincide with the center of mass of the body that it is modeling.

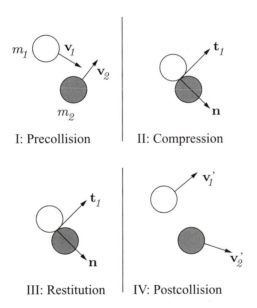

I: Precollision   II: Compression

III: Restitution   IV: Postcollision

FIGURE 6.3. The four stages of collision

As summarized in Figure 6.3, we have four time periods to examine: I, just prior to impact; II, during the compression phase of the impact; III, during the restitution phase of the impact; and IV, immediately after the impact. This figure also summarizes the notation for the velocities that we will use.

When the two bodies are in contact they are assumed to have a common unit normal $\mathbf{n}$ at the single point of contact. This vector is normal to the lateral surfaces of both bodies at the unique point of contact. We also define a set of orthonormal vectors $\{\mathbf{n}, \mathbf{t}_1, \mathbf{t}_2\}$ by selecting $\mathbf{t}_1$ and $\mathbf{t}_2$ to be unit tangent vectors to the lateral surfaces of both bodies at the point of contact. The impact is assumed to be such that these vectors are constant for the duration of the collision.[4]

The impact is assumed to occur at time $t = t_0$. Stage II of the collision occurs during the time interval (t_0, t_1), while stage III occurs during the time interval $[t_1, t_2)$. At $t = t_2$, the bodies have just lost contact. At the end of stage II, the velocities of the bodies in the direction of $\mathbf{n}$ are assumed to be equal $(= v_{II})$.

6.3.2   Linear Impulses During Impact

Pertaining to the forces exerted by the bodies on each other during the impact, let $\mathbf{F}_{1d}$ and $\mathbf{F}_{1r}$ be the forces exerted by body 2 on body 1 during stages II and III, respectively. Similarly, we denote by $\mathbf{F}_{2d}$ and $\mathbf{F}_{2r}$ the forces exerted by body 1 on body 2 during stages II and III, respectively. All other forces acting on bodies 1 and 2 are assumed to have the resultants $\mathbf{R}_1$ and $\mathbf{R}_2$, respectively.

The following 3 assumptions are made for the aforementioned forces. First, the impact is frictionless:

$$\mathbf{F}_{1d} = F_{1d}\mathbf{n}, \quad \mathbf{F}_{1r} = F_{1r}\mathbf{n}, \quad \mathbf{F}_{2d} = F_{2d}\mathbf{n}, \quad \mathbf{F}_{2r} = F_{2r}\mathbf{n}.$$

That is, these forces have no tangential components. Second, the linear impulse of these forces dominates those due to $\mathbf{R}_1$ and $\mathbf{R}_2$:

$$\int_{t_0}^{t_1} \mathbf{F}_{1d}(\tau) + \mathbf{R}_1(\tau)d\tau \;\approx\; \int_{t_0}^{t_1} \mathbf{F}_{1d}(\tau)d\tau,$$

$$\int_{t_1}^{t_2} \mathbf{F}_{1r}(\tau) + \mathbf{R}_1(\tau)d\tau \;\approx\; \int_{t_1}^{t_2} \mathbf{F}_{1r}(\tau)d\tau,$$

$$\int_{t_0}^{t_1} \mathbf{F}_{2d}(\tau) + \mathbf{R}_2(\tau)d\tau \;\approx\; \int_{t_0}^{t_1} \mathbf{F}_{2d}(\tau)d\tau,$$

$$\int_{t_1}^{t_2} \mathbf{F}_{2r}(\tau) + \mathbf{R}_2(\tau)d\tau \;\approx\; \int_{t_1}^{t_2} \mathbf{F}_{2r}(\tau)d\tau.$$

It is normally assumed that the time interval $[t_0, t_2]$ is small for this assumption to hold. Finally, we assume equal and opposite collisional forces: $\mathbf{F}_{1r} = -\mathbf{F}_{2r}$ and $\mathbf{F}_{1d} = -\mathbf{F}_{2d}$.

In addition, one defines the coefficient of restitution e:

$$e = \frac{\int_{t_1}^{t_2} \mathbf{F}_{1r}(\tau) \cdot \mathbf{n}d\tau}{\int_{t_0}^{t_1} \mathbf{F}_{1d}(\tau) \cdot \mathbf{n}d\tau} = \frac{\int_{t_1}^{t_2} \mathbf{F}_{2r}(\tau) \cdot \mathbf{n}d\tau}{\int_{t_0}^{t_1} \mathbf{F}_{2d}(\tau) \cdot \mathbf{n}d\tau}.$$

---

[4]An example illustrating these three vectors is shown in Figure 6.6 below. We also note that for many problems these vectors will coincide with the Cartesian basis vectors.

Here, we used the equal and opposite nature of the interaction forces. You should notice that if $e = 1$, then the linear impulse during the compression stage is equal to the linear impulse during the restitution phase. In this case, the collision is said to be perfectly elastic. If $e = 0$, then the linear impulse during the restitution phase is zero, and the collision is said to be perfectly plastic. In general, $0 \le e \le 1$, and e must be determined from an experiment.

To write the coefficient of restitution in a more convenient form using velocities, we first record the following integral forms of the balance of linear momentum for each particle in the direction of $\mathbf{n}$ during stages II and III:

$$m_1 v_{II} - m_1 \mathbf{v}_1 \cdot \mathbf{n} = \int_{t_0}^{t_1} \mathbf{F}_{1d}(\tau) \cdot \mathbf{n} d\tau,$$

$$m_2 v_{II} - m_2 \mathbf{v}_2 \cdot \mathbf{n} = \int_{t_0}^{t_1} \mathbf{F}_{2d}(\tau) \cdot \mathbf{n} d\tau,$$

$$m_1 \mathbf{v}_1' \cdot \mathbf{n} - m_1 v_{II} = \int_{t_1}^{t_2} \mathbf{F}_{1r}(\tau) \cdot \mathbf{n} d\tau = e \int_{t_0}^{t_1} \mathbf{F}_{1d}(\tau) \cdot \mathbf{n} d\tau,$$

$$m_2 \mathbf{v}_2' \cdot \mathbf{n} - m_2 v_{II} = \int_{t_1}^{t_2} \mathbf{F}_{2r}(\tau) \cdot \mathbf{n} d\tau = e \int_{t_0}^{t_1} \mathbf{F}_{2d}(\tau) \cdot \mathbf{n} d\tau.$$

In these four equations v_{II} is the common velocity in the direction of $\mathbf{n}$ at the end of the compression phase. From these four equations, we can determine this velocity:

$$v_{II} = \frac{\mathbf{v}_1' \cdot \mathbf{n} + e \mathbf{v}_1 \cdot \mathbf{n}}{1 + e} = \frac{\mathbf{v}_2' \cdot \mathbf{n} + e \mathbf{v}_2 \cdot \mathbf{n}}{1 + e}.$$

We can also manipulate these four equations to find a familiar expression for the coefficient of restitution:

$$e = \frac{\mathbf{v}_2' \cdot \mathbf{n} - \mathbf{v}_1' \cdot \mathbf{n}}{\mathbf{v}_1 \cdot \mathbf{n} - \mathbf{v}_2 \cdot \mathbf{n}}.$$

This equation is used by many authors as the definition of the coefficient of restitution.

6.3.3 Linear Momenta

We now consider the integral forms of the balance of linear momentum for each particle. The time interval of integration is the duration of the impact:

$$m_1 \mathbf{v}_1' - m_1 \mathbf{v}_1 = \int_{t_0}^{t_1} \mathbf{F}_{1d}(\tau) + \mathbf{R}_1(\tau) d\tau + \int_{t_1}^{t_2} \mathbf{F}_{1r}(\tau) + \mathbf{R}_1(\tau) d\tau,$$

$$m_2 \mathbf{v}_2' - m_2 \mathbf{v}_2 = \int_{t_0}^{t_1} \mathbf{F}_{2d}(\tau) + \mathbf{R}_2(\tau) d\tau + \int_{t_1}^{t_2} \mathbf{F}_{2r}(\tau) + \mathbf{R}_2(\tau) d\tau.$$

Using the coefficient of restitution e, the assumptions that $\mathbf{F}_{1d} = -\mathbf{F}_{2d}$, $\mathbf{F}_{1r} = -\mathbf{F}_{2r}$, and that the linear impulses of these forces dominate those due to $\mathbf{R}_1$ and $\mathbf{R}_2$, we find that

$$m_1\mathbf{v}_1' - m_1\mathbf{v}_1 = (1+e)\int_{t_0}^{t_1} \mathbf{F}_{1d}(\tau)d\tau,$$

$$m_2\mathbf{v}_2' - m_2\mathbf{v}_2 = -(1+e)\int_{t_0}^{t_1} \mathbf{F}_{1d}(\tau)d\tau.$$

We now take the three components of these equations with respect to the basis vectors $\{\mathbf{n}, \mathbf{t}_1, \mathbf{t}_2\}$. The components of these equations in the tangential directions imply that the linear momenta of the particles in these directions are conserved:

$$\mathbf{v}_1' \cdot \mathbf{t}_1 = \mathbf{v}_1 \cdot \mathbf{t}_1, \quad \mathbf{v}_1' \cdot \mathbf{t}_2 = \mathbf{v}_1 \cdot \mathbf{t}_2,$$

$$\mathbf{v}_2' \cdot \mathbf{t}_1 = \mathbf{v}_2 \cdot \mathbf{t}_1, \quad \mathbf{v}_2' \cdot \mathbf{t}_2 = \mathbf{v}_2 \cdot \mathbf{t}_2.$$

In addition, from the two components in the $\mathbf{n}$ direction, we find that the linear momentum of the system in this direction is conserved:

$$m_2\mathbf{v}_2' \cdot \mathbf{n} + m_1\mathbf{v}_1' \cdot \mathbf{n} = m_2\mathbf{v}_2 \cdot \mathbf{n} + m_1\mathbf{v}_1 \cdot \mathbf{n},$$

$$m_2\mathbf{v}_2' \cdot \mathbf{n} - m_2\mathbf{v}_2 \cdot \mathbf{n} = -(1+e)\int_{t_0}^{t_1} \mathbf{F}_{1d}(\tau) \cdot \mathbf{n}d\tau.$$

The previous 6 equations should be sufficient to determine the 6 postimpact velocities $\mathbf{v}_1'$ and $\mathbf{v}_2'$ provided that one knows the preimpact velocities and the linear impulse of $\mathbf{F}_{1d}$ during the collision. However, this linear impulse is problem-dependent, and so one generally supplements these equations with the specification of the coefficient of restitution to determine the postimpact velocity vectors.

6.3.4   The Postimpact Velocities

It is convenient at this point to summarize the equations and show how they are used to solve certain problems. For the problems of interest one is often given e, $\mathbf{v}_1$, $\mathbf{v}_2$, m_1, m_2, and $\{\mathbf{n}, \mathbf{t}_1, \mathbf{t}_2\}$, and asked to calculate the postimpact velocity vectors $\mathbf{v}_1'$ and $\mathbf{v}_2'$.

From the previous 2 sets of equations and the specification of the coefficient of restitution, we see that we have 1 equation to determine the deformational linear impulse and 6 equations to determine the postimpact velocity vectors. With some algebra, one obtains

$$\mathbf{v}_1' = (\mathbf{v}_1 \cdot \mathbf{t}_1)\,\mathbf{t}_1 + (\mathbf{v}_1 \cdot \mathbf{t}_2)\,\mathbf{t}_2$$

$$+ \frac{1}{m_1 + m_2}\left((m_1 - em_2)\mathbf{v}_1 \cdot \mathbf{n} + (1+e)m_2\mathbf{v}_2 \cdot \mathbf{n}\right)\mathbf{n},$$

$$\mathbf{v}_2' = (\mathbf{v}_2 \cdot \mathbf{t}_1)\,\mathbf{t}_1 + (\mathbf{v}_2 \cdot \mathbf{t}_2)\,\mathbf{t}_2$$
$$+ \frac{1}{m_1 + m_2}\left((m_2 - em_1)\mathbf{v}_2 \cdot \mathbf{n} + (1+e)m_1\mathbf{v}_1 \cdot \mathbf{n}\right)\mathbf{n}.$$

You should notice how the postimpact velocity vectors depend on the mass ratios. We could also calculate the linear impulse of $\mathbf{F}_{1d}$, but we do not pause to do so here.

It is important to remember that the above expressions for the postimpact velocities are consequences of the following: (i) Linear momentum of each particle in the tangent directions is conserved during the impact, (ii) the combined linear momentum of the system in the normal direction is conserved during the impact, and (iii) the coefficient of restitution needs to be provided to determine the aforementioned velocity vectors.[5]

6.3.5 Kinetic Energy and the Coefficient of Restitution

Previously, we have used two equivalent definitions of the coefficient of restitution:
$$e = \frac{\mathbf{v}_2' \cdot \mathbf{n} - \mathbf{v}_1' \cdot \mathbf{n}}{\mathbf{v}_1 \cdot \mathbf{n} - \mathbf{v}_2 \cdot \mathbf{n}},$$
and
$$e = \frac{\int_{t_1}^{t_2} \mathbf{F}_{1r}(\tau) \cdot \mathbf{n} d\tau}{\int_{t_0}^{t_1} \mathbf{F}_{1d}(\tau) \cdot \mathbf{n} d\tau} = \frac{\int_{t_1}^{t_2} \mathbf{F}_{2r}(\tau) \cdot \mathbf{n} d\tau}{\int_{t_0}^{t_1} \mathbf{F}_{2d}(\tau) \cdot \mathbf{n} d\tau}.$$
Often, the change in kinetic energy is used to specify the coefficient of restitution. We now examine why this is possible.

The kinetic energy of the system just prior to impact T and the kinetic energy immediately following the collision T' are, by definition,
$$T = \frac{1}{2}m_1\mathbf{v}_1 \cdot \mathbf{v}_1 + \frac{1}{2}m_2\mathbf{v}_2 \cdot \mathbf{v}_2, \quad T' = \frac{1}{2}m_1\mathbf{v}_1' \cdot \mathbf{v}_1' + \frac{1}{2}m_2\mathbf{v}_2' \cdot \mathbf{v}_2'.$$
We recall that the collision changes only the $\mathbf{n}$ components of the velocity vectors. Hence,
$$T - T' = \frac{1}{2}m_1\left((\mathbf{v}_1 \cdot \mathbf{n})^2 - (\mathbf{v}_1' \cdot \mathbf{n})^2\right) + \frac{1}{2}m_2\left((\mathbf{v}_2 \cdot \mathbf{n})^2 - (\mathbf{v}_2' \cdot \mathbf{n})^2\right).$$
Substituting for the normal components of the postimpact velocity vectors, we obtain the well-known equation
$$T - T' = \frac{m_1 m_2}{2m_1 + 2m_2}\left(\mathbf{v}_1 \cdot \mathbf{n} - \mathbf{v}_2 \cdot \mathbf{n}\right)^2\left(1 - e^2\right).$$

---

[5]It should be clear to you that, given the preimpact velocity vectors, the balance laws give only 6 equations from which one needs to determine the 6 postimpact velocities and the linear impulses of $\mathbf{F}_{1d}$ and $\mathbf{F}_{1r}$. The introduction of the coefficient of restitution e and the assumption of a common normal velocity v_{II} at time $t = t_1$ gives 2 more equations which renders the system of equations solvable, i.e., these 2 equations close the system of equations.

Hence, if $-1 \le e \le 1$, the kinetic energy of the system cannot be increased as a result of the impact. One can also invert this equation to obtain a definition of e in terms of the postimpact velocity vectors and the change in kinetic energy.

6.3.6   Negative Values of the Coefficient of Restitution

It follows from the previous equation that assuming that e is negative does not preclude energy loss during an impact. Indeed, Brach [10] considers the example of a ball shattering and then passing through a window as an example of a problem where e is negative. In problems where e is negative,

$$ e = \frac{\mathbf{v}_2' \cdot \mathbf{n} - \mathbf{v}_1' \cdot \mathbf{n}}{\mathbf{v}_1 \cdot \mathbf{n} - \mathbf{v}_2 \cdot \mathbf{n}} < 0 \,, $$

and hence, $\mathbf{v}_1' \cdot \mathbf{n} - \mathbf{v}_2' \cdot \mathbf{n}$ and $\mathbf{v}_1 \cdot \mathbf{n} - \mathbf{v}_2 \cdot \mathbf{n}$ have the same sign. As a result, the colliding bodies must interpenetrate and pass through each other during the course of the impact. To eliminate this behavior, it is generally assumed that e is positive.

6.4   Impact of a Particle and a Massive Object

To illustrate the previous results, consider a particle of mass m_1 that collides with a massive object (see Figure 6.4). We expect that the velocity of the massive object will not be affected by the collision.

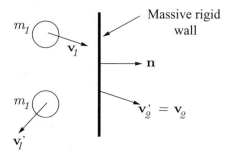

FIGURE 6.4. Impact of a particle and a massive object

For the problem at hand, m_2 is far greater that m_1:

$$ \frac{m_1}{m_1 + m_2} \approx 0 \,, \qquad \frac{m_2}{m_1 + m_2} \approx 1 \,. $$

Substituting these results into the expressions for the postimpact velocities, we find that

$$ \mathbf{v}_1' \;=\; (\mathbf{v}_1 \cdot \mathbf{t}_1)\,\mathbf{t}_1 + (\mathbf{v}_1 \cdot \mathbf{t}_2)\,\mathbf{t}_2 + (-e\mathbf{v}_1 \cdot \mathbf{n} + (1 + e)\mathbf{v}_2 \cdot \mathbf{n})\,\mathbf{n} \,, $$

$$\mathbf{v}_2' = (\mathbf{v}_2 \cdot \mathbf{t}_1)\,\mathbf{t}_1 + (\mathbf{v}_2 \cdot \mathbf{t}_2)\,\mathbf{t}_2 + (\mathbf{v}_2 \cdot \mathbf{n})\,\mathbf{n} = \mathbf{v}_2 \,.$$

As expected, $\mathbf{v}_2$ was unaltered by the collision. You should also notice that if $e = 1$ and $\mathbf{v}_2 = \mathbf{0}$, then the particle rebounds with its velocity in the direction of $\mathbf{n}$ reversed, as expected. Finally, for a plastic collision, $e = 0$ and the velocity of m_1 in the direction of $\mathbf{n}$ attains the velocity of the massive object in this direction.

6.5   Collision of Two Spheres

Another example of interest is shown in Figure 6.5. There, a sphere of mass m_1 and radius R that is moving at a constant velocity $\mathbf{v}_1 = 100\mathbf{E}_x$ (m/sec) collides with a stationary sphere of radius r and mass $m_2 = 2m_1$.

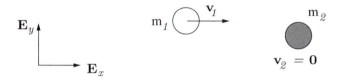

FIGURE 6.5. An impact of two spheres

At the instant of impact, the position vectors of the centers of mass of the spheres differ in height by an amount h. Given that the coefficient of restitution is $e = 0.5$, one seeks to determine the velocity vectors of the spheres immediately following the impact.

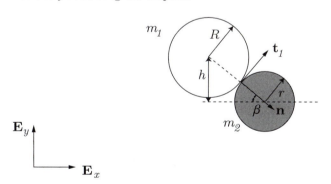

FIGURE 6.6. The geometry of the impact

It is first necessary to determine the normal and tangent vectors at the contact point of the spheres during the impact. Referring to Figure 6.6, we see that these vectors are

$$\mathbf{n} = \cos(\beta)\mathbf{E}_x - \sin(\beta)\mathbf{E}_y\,, \quad \mathbf{t}_1 = \cos(\beta)\mathbf{E}_y + \sin(\beta)\mathbf{E}_x\,, \quad \mathbf{t}_2 = \mathbf{E}_z\,.$$

Here, the angle β is defined by the relations

$$\sin(\beta) = \frac{h}{R+r}, \quad \cos(\beta) = \frac{\sqrt{(r+R)^2 - h^2}}{r+R}.$$

Notice that this angle depends on h, r, and R.

One method of solving this problem is to use the formulae given previously for the postimpact velocity vectors:

$$\begin{aligned}
\mathbf{v}_1' &= (\mathbf{v}_1 \cdot \mathbf{t}_1)\,\mathbf{t}_1 + (\mathbf{v}_1 \cdot \mathbf{t}_2)\,\mathbf{t}_2 \\
&\quad + \frac{1}{m_1 + m_2}\left((m_1 - em_2)\mathbf{v}_1 \cdot \mathbf{n} + (1+e)m_2\mathbf{v}_2 \cdot \mathbf{n}\right)\mathbf{n}, \\
\mathbf{v}_2' &= (\mathbf{v}_2 \cdot \mathbf{t}_1)\,\mathbf{t}_1 + (\mathbf{v}_2 \cdot \mathbf{t}_2)\,\mathbf{t}_2 \\
&\quad + \frac{1}{m_1 + m_2}\left((m_2 - em_1)\mathbf{v}_2 \cdot \mathbf{n} + (1+e)m_1\mathbf{v}_1 \cdot \mathbf{n}\right)\mathbf{n}.
\end{aligned}$$

Substituting for the values for the problem at hand,

$$\mathbf{v}_1 = 100\mathbf{E}_x, \quad \mathbf{v}_2 = \mathbf{0}, \quad e = 0.5, \quad m_2 = 2m_1,$$

into these equations, we find that

$$\begin{aligned}
\mathbf{v}_1' &= (\mathbf{v}_1 \cdot \mathbf{t}_1)\,\mathbf{t}_1 = 100\sin(\beta)\mathbf{t}_1, \\
\mathbf{v}_2' &= \frac{1}{2}(\mathbf{v}_1 \cdot \mathbf{n})\,\mathbf{n} = 50\cos(\beta)\mathbf{n}.
\end{aligned}$$

Notice that the initial velocity of the sphere of radius R in the direction of $\mathbf{n}$ has been annihilated by the collision.

6.6 Summary

The first new concept in this chapter was the definition of the linear momentum $\mathbf{G} = m\mathbf{v}$. In addition, the integral form of the balance of linear momentum was introduced:

$$\mathbf{G}(t_1) - \mathbf{G}(t_0) = \int_{t_0}^{t_1} \mathbf{F}\,dt.$$

This equation is assumed to hold for all intervals of time, and, hence, for all times t_0 and t_1. Because this balance law does not assume that $\mathbf{v}$ is differentiable, it more general than $\mathbf{F} = m\mathbf{a}$.

The angular momentum $\mathbf{H}_O = \mathbf{r} \times \mathbf{G}$ of a particle relative to a fixed point O was introduced in Section 6.2. The time-rate of change of this momentum can be determined using the angular momentum theorem:

$$\dot{\mathbf{H}}_O = \mathbf{r} \times \mathbf{F}.$$

Associated with $\mathbf{G}$ and $\mathbf{H}_O$ are instances where certain components of these vectors are conserved. For a given vector $\mathbf{c}$, the conservations of $\mathbf{G} \cdot \mathbf{c}$ and $\mathbf{H}_O \cdot \mathbf{c}$ are established using the following equations:

$$\frac{d}{dt}\left(\mathbf{G} \cdot \mathbf{c}\right) = \mathbf{F} \cdot \mathbf{c} + \mathbf{G} \cdot \dot{\mathbf{c}} = 0,$$

$$\frac{d}{dt}\left(\mathbf{H}_O \cdot \mathbf{c}\right) = (\mathbf{r} \times \mathbf{F}) \cdot \mathbf{c} + \mathbf{H}_O \cdot \dot{\mathbf{c}} = 0.$$

Several examples of linear and angular momenta conservations were discussed in the chapter. Specifically, the conservation of a component of $\mathbf{G}$ for impact problems and projectile problems was covered. The more conceptually challenging conservation of a component of $\mathbf{H}_O$ was illustrated using a central force problem, Kepler's problem, and a particle moving on the surface of a smooth cone.

The majority of the chapter was devoted to a discussion of impact problems. Specifically, given two bodies of masses m_1 and m_2 whose respective postimpact velocity vectors are $\mathbf{v}_1$ and $\mathbf{v}_2$, we wished to find the postimpact velocity vectors $\mathbf{v}_1'$ and $\mathbf{v}_2'$. The collisions of interest were restricted to cases where there was a unique point of contact with a well-defined unit normal vector $\mathbf{n}$. This normal vector was then used to construct a right-handed orthonormal triad $\{\mathbf{n}, \mathbf{t}_1, \mathbf{t}_2\}$. To solve the problems of interest, it was necessary to introduce a phenomenological constant, the coefficient of restitution e. In Sections 3.2 and 3.5, three alternative definitions of this constant were presented:

$$e = \frac{\mathbf{v}_2' \cdot \mathbf{n} - \mathbf{v}_1' \cdot \mathbf{n}}{\mathbf{v}_1 \cdot \mathbf{n} - \mathbf{v}_2 \cdot \mathbf{n}},$$

$$e = \frac{\int_{t_1}^{t_2} \mathbf{F}_{1r}(\tau) \cdot \mathbf{n} d\tau}{\int_{t_0}^{t_1} \mathbf{F}_{1d}(\tau) \cdot \mathbf{n} d\tau} = \frac{\int_{t_1}^{t_2} \mathbf{F}_{2r}(\tau) \cdot \mathbf{n} d\tau}{\int_{t_0}^{t_1} \mathbf{F}_{2d}(\tau) \cdot \mathbf{n} d\tau},$$

$$T - T' = \frac{m_1 m_2}{2m_1 + 2m_2}(\mathbf{v}_1 \cdot \mathbf{n} - \mathbf{v}_2 \cdot \mathbf{n})^2 \left(1 - e^2\right).$$

We also noted that the restriction $0 \le e \le 1$ is normally imposed so as to preclude interpenetrability of the impacting bodies. If $e = 1$, the collision is said to be perfectly elastic and if $e = 0$, then the collision is said to be perfectly plastic.

Using the definition of the coefficient of restitution, conservation of linear momenta of each particle in the $\mathbf{t}_1$ and $\mathbf{t}_2$ directions, and conservation of the total linear momentum of the system of particles in the $\mathbf{n}$ direction, the following results were obtained for the postimpact velocity vectors:

$$\mathbf{v}_1' = (\mathbf{v}_1 \cdot \mathbf{t}_1)\,\mathbf{t}_1 + (\mathbf{v}_1 \cdot \mathbf{t}_2)\,\mathbf{t}_2$$
$$+ \frac{1}{m_1 + m_2}\left((m_1 - em_2)\mathbf{v}_1 \cdot \mathbf{n} + (1 + e)m_2\mathbf{v}_2 \cdot \mathbf{n}\right)\mathbf{n},$$

$$\mathbf{v}_2' = (\mathbf{v}_2 \cdot \mathbf{t}_1)\,\mathbf{t}_1 + (\mathbf{v}_2 \cdot \mathbf{t}_2)\,\mathbf{t}_2$$

$$+ \frac{1}{m_1 + m_2} \left((m_2 - em_1)\mathbf{v}_2 \cdot \mathbf{n} + (1 + e)m_1\mathbf{v}_1 \cdot \mathbf{n} \right) \mathbf{n} \,.$$

The solution of impact problems using these formulae were presented in Sections 4 and 5.

6.7 Exercises

The following short exercises are intended to assist you in reviewing Chapter 6.

6.1 A particle of mass m is in motion on a smooth horizontal surface. Using $\mathbf{F} = m\mathbf{a}$, show that the resultant force $\mathbf{F}$ acting on the particle is zero, and hence its linear momentum $\mathbf{G}$ remains constant.

6.2 For a particle of mass m which is falling under the influence of gravitational force $-mg\mathbf{E}_z$, show that $\mathbf{G} \cdot \mathbf{E}_z$ is not conserved.

6.3 A particle of mass m is in motion on a smooth horizontal surface. Here, a gravitational force $-mg\mathbf{E}_z$ and an applied force $\mathbf{P} = P(t)\mathbf{E}_x$ acts on the particle. Using $\mathbf{F} = m\mathbf{a}$, show that $\mathbf{F} = \mathbf{P}$. Furthermore, show that

$$\mathbf{G}(t) = \mathbf{G}(t_0) + \left(\int_{t_0}^{t} P(\tau)d\tau \right) \mathbf{E}_x \,.$$

6.4 Starting from $\mathbf{H}_O = \mathbf{r} \times m\mathbf{v}$, prove the angular momentum theorem: $\dot{\mathbf{H}}_O = \mathbf{r} \times \mathbf{F}$.

6.5 For a particle of mass m moving on a horizontal plane $z = 0$, show that $\mathbf{H}_O = mr^2\dot{\theta}\mathbf{E}_z$. When this momentum is conserved, show, with the aid of graphs of $\dot{\theta}$ as a function of r for various values of $h = \mathbf{H}_O \cdot \mathbf{E}_z$, that the sign of $\dot{\theta}$ cannot change. What does this imply about the motion of the particle?

6.6 Consider the example discussed in Section 2.5. Show that $\mathbf{H}_O \cdot \mathbf{E}_z$ is conserved when the spring is replaced by an inextensible string of length L. In your solution, assume that the string is being fed from an eyelet at O. Consequently, the length of string between O and the particle can change: $L = L(t)$.

6.7 Does the solution to Exercise 6.6 change if the angle $\alpha = 0$? What physical problem does this case correspond to?

6.8 Recall Kepler's problem discussed in Section 2.4. Show that angular momentum conservation can be used to reduce the equations governing θ and r to a single differential equation:

$$\ddot{r} - \frac{h^2}{m^2 r^3} + \frac{GMm}{r^2} = 0 \,.$$

Show that this equation predicts a circular motion $r = r_0$, where

$$r_0 = \frac{h^2}{GMm^3}, \qquad \dot{\theta} = \frac{h}{mr_0^2}.$$

Finally, show that the velocity vector of the particle of mass m must be

$$\mathbf{v} = \frac{GMm^2}{h}\mathbf{e}_\theta$$

6.9 Using the following result, which was discussed in Section 3.5,

$$T - T' = \frac{m_1 m_2}{2m_1 + 2m_2}\left(\mathbf{v}_1 \cdot \mathbf{n} - \mathbf{v}_2 \cdot \mathbf{n}\right)^2 \left(1 - e^2\right),$$

under which circumstances is the kinetic energy loss maximized or minimized in a collision?

6.10 Why is it necessary to know the coefficient of restitution e in order to solve an impact problem?

6.11 When two bodies are in contact at a point, there are two possible choices for $\mathbf{n}$: $\pm\mathbf{n}$. Why do the formulae for e, $\mathbf{v}_1'$, and $\mathbf{v}_2'$, that are given in the summary section, give the same results for either choice of $\mathbf{n}$?

6.12 Recall the problem discussed in Section 4 where a particle of mass m impacts a massive object. If $e = 0$, verify that $\mathbf{v}_1' \cdot \mathbf{n} = \mathbf{v}_2 \cdot \mathbf{n}$. On physical grounds why are the components of $\mathbf{v}_1$ in the $\mathbf{t}_1$ and $\mathbf{t}_2$ directions unaltered by the collision?

6.13 Using the results for the problem discussed in Section 4, show that if e is negative, then the particle passes through the massive object. When $e = -1$, show that the particle is unaffected by its collision with the massive object.

6.14 Using the results of the example discussed in Section 5, determine the postimpact velocity vectors of the spheres for the two cases $h = 0$ and $h = R + r$.

7
Systems of Particles

TOPICS

Here, we extend several results pertaining to a single particle to a system of particles. We start by defining the linear momentum, angular momenta, and kinetic energy for a system of particles. Next, we introduce a new concept, the center of mass C of a system of particles. A discussion of the conservation of kinematical quantities follows, which we illustrate with two detailed examples.

7.1 Preliminaries

We consider a system of n particles, each of mass m_i $(i = 1, \ldots, n)$. The position vector of the particle of mass m_i relative to a fixed point O is denoted by $\mathbf{r}_i$. Several quantities pertaining to the kinematics of this system are shown in Figure 7.1.

The velocity vector of the particle of mass m_i is defined as

$$\mathbf{v}_i = \frac{d\mathbf{r}_i}{dt},$$

while the (absolute) acceleration vector $\mathbf{a}_i$ of this particle is

$$\mathbf{a}_i = \frac{d^2\mathbf{r}_i}{dt^2} = \frac{d\mathbf{v}_i}{dt}.$$

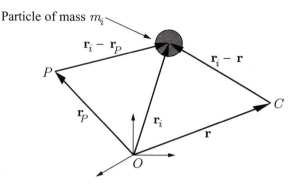

FIGURE 7.1. Some kinematical quantities

We then have that the linear momentum $\mathbf{G}_i$ of the particle of mass m_i is

$$\mathbf{G}_i = m_i \mathbf{v}_i \,,$$

and the angular momentum $\mathbf{H}_{Pi}$ of the particle of mass m_i relative to a point P is

$$\mathbf{H}_{Pi} = (\mathbf{r}_i - \mathbf{r}_P) \times \mathbf{G}_i \,,$$

where $\mathbf{r}_P$ is the position vector of the point P relative to O. Finally, the kinetic energy T_i of the particle of mass m_i is

$$T_i = \frac{1}{2} m_i \mathbf{v}_i \cdot \mathbf{v}_i \,.$$

7.2   Center of Mass, Momenta, and Kinetic Energy

7.2.1   The Center of Mass

The center of mass C of the system of particles is defined as the point whose position vector $\mathbf{r}$ is defined by

$$\mathbf{r} = \frac{1}{m} \sum_{k=1}^{n} m_k \mathbf{r}_k \,,$$

where

$$m = \sum_{k=1}^{n} m_k$$

is the total mass of the system of particles.

The velocity $\mathbf{v}$ of the center of mass is obtained from the above equation by differentiating it with respect to time:

$$\mathbf{v} = \frac{1}{m} \sum_{k=1}^{n} m_k \mathbf{v}_k = \frac{1}{m} \sum_{k=1}^{n} \mathbf{G}_k \,.$$

Hence, the velocity of the center of mass is a weighted sum of the velocities of the particles.

It is convenient to record the following identities:

$$\sum_{k=1}^{n} m_k(\mathbf{r} - \mathbf{r}_k) = \mathbf{0}, \quad \sum_{k=1}^{n} m_k(\mathbf{v} - \mathbf{v}_k) = \mathbf{0}.$$

These identities will shortly be used to derive convenient expressions for the linear and angular momenta and kinetic energy of a system of particles. The method of manipulation used there is similar to that which will be used later for rigid bodies.

7.2.2 Linear Momentum

The linear momentum $\mathbf{G}$ of the system of particles is the sum of the linear momenta of the individual particles. It follows from the definition of the center of mass that

$$\mathbf{G} = m\frac{d\mathbf{r}}{dt} = \sum_{k=1}^{n} m_k\frac{d\mathbf{r}_k}{dt} = \sum_{k=1}^{n} \mathbf{G}_k.$$

In words, the linear momentum of the center of mass is the linear momentum of the system.

7.2.3 Angular Momentum

Similarly, the angular momentum $\mathbf{H}_P$ of the system of particles relative to a point P, whose position vector relative to O is $\mathbf{r}_P$, is the sum of the individual angular momenta:

$$\mathbf{H}_P = \sum_{k=1}^{n} \mathbf{H}_{Pk} = \sum_{k=1}^{n} (\mathbf{r}_k - \mathbf{r}_P) \times m_k \mathbf{v}_k.$$

Using the definition of the center of mass, we can write $\mathbf{H}_P$ in a more convenient form:

$$\begin{aligned}
\mathbf{H}_P &= \sum_{k=1}^{n} \mathbf{H}_{Pk} = \sum_{k=1}^{n} (\mathbf{r}_k - \mathbf{r}_P) \times \mathbf{G}_k \\
&= \sum_{k=1}^{n} (\mathbf{r}_k - \mathbf{r} + \mathbf{r} - \mathbf{r}_P) \times \mathbf{G}_k \\
&= \sum_{k=1}^{n} (\mathbf{r}_k - \mathbf{r}) \times \mathbf{G}_k + (\mathbf{r} - \mathbf{r}_P) \times \sum_{k=1}^{n} \mathbf{G}_k.
\end{aligned}$$

That is,

$$\mathbf{H}_P = \mathbf{H}_C + (\mathbf{r} - \mathbf{r}_P) \times \mathbf{G}.$$

In this equation,

$$\mathbf{H}_C = \sum_{k=1}^{n} (\mathbf{r}_k - \mathbf{r}) \times m_k \mathbf{v}_k$$

is the angular momentum of the system of particles relative to its center of mass C.

7.2.4 Kinetic Energy

The kinetic energy T of the system of particles is defined to be the sum of their individual kinetic energies:

$$T = \sum_{k=1}^{n} T_k = \sum_{k=1}^{n} \frac{1}{2} m_k \mathbf{v}_k \cdot \mathbf{v}_k \,.$$

In general, the kinetic energy of the system is not equal to the kinetic energy of its center of mass. This shall shortly become evident.

Using the center of mass, T can be expressed in another form:

$$
\begin{aligned}
T &= \sum_{k=1}^{n} T_k = \frac{1}{2} \sum_{k=1}^{n} m_k \mathbf{v}_k \cdot \mathbf{v}_k \\
&= \frac{1}{2} \sum_{k=1}^{n} m_k \mathbf{v}_k \cdot \mathbf{v}_k - \mathbf{v} \cdot \sum_{k=1}^{n} m_k (\mathbf{v}_k - \mathbf{v}) \,.
\end{aligned}
$$

Notice that we have added a term to the right-hand side that is equal to zero. Some minor manipulations of this result yields

$$
\begin{aligned}
T &= \frac{1}{2} \sum_{k=1}^{n} m_k \mathbf{v}_k \cdot \mathbf{v}_k - \mathbf{v} \cdot \sum_{k=1}^{n} m_k (\mathbf{v}_k - \mathbf{v}) \\
&= \frac{1}{2} \sum_{k=1}^{n} m_k \mathbf{v} \cdot \mathbf{v} + \frac{1}{2} \sum_{k=1}^{n} m_k (\mathbf{v}_k \cdot \mathbf{v}_k - 2\mathbf{v}_k \cdot \mathbf{v} + \mathbf{v} \cdot \mathbf{v}) \,.
\end{aligned}
$$

Completing the square, we obtain the final desired result:

$$T = \frac{1}{2} m \mathbf{v} \cdot \mathbf{v} + \frac{1}{2} \sum_{k=1}^{n} m_k (\mathbf{v}_k - \mathbf{v}) \cdot (\mathbf{v}_k - \mathbf{v}) \,.$$

That is, the kinetic energy of a system of particles is the kinetic energy of its center of mass plus another term which is proportional to the magnitude of the velocity of the particles relative to the center of mass.[1]

---

[1] This result is sometimes known as the Koenig decomposition of the kinetic energy of a system of particles. In Chapter 9, the corresponding decomposition for a rigid body is discussed.

7.2.5   Kinetics of Systems of Particles

For each individual particle one has Euler's first law (Newton's second law or the balance of linear momentum):

$$\mathbf{F}_i = m_i \mathbf{a}_i \ .$$

Adding these n equations and using the definition of the center of mass, we find that

$$\mathbf{F} = m\mathbf{a} \ ,$$

where $\mathbf{F}$ is the resultant force acting on the system of particles:

$$\mathbf{F} = \sum_{k=1}^{n} \mathbf{F}_k \ .$$

The equation $\mathbf{F} = m\mathbf{a}$ is extremely useful. One uses it to solve for the motion of the center of mass of the system.

In many systems of particles problems, determining the motions of the particles, by solving the set of coupled second-order ordinary differential equations for $\mathbf{r}_i(t)$, is an extremely difficult task;[2] one that is well beyond the scope of an undergraduate engineering dynamics course.

7.3   Conservation of Linear Momentum

We first consider conditions for the conservation of the component of the linear momentum $\mathbf{G}$ in the direction of a given vector $\mathbf{c} = \mathbf{c}(t)$. The result parallels the case for a single particle discussed in Section 1 of Chapter 6, and so we merely quote it: If $\mathbf{F} \cdot \mathbf{c} + \mathbf{G} \cdot \mathbf{c} = 0$, then $\mathbf{G} \cdot \mathbf{c}$ is conserved.

7.3.1   The Cart and the Pendulum

As a first example of the conservation of linear momentum, consider the system shown in Figure 7.2. A particle of mass m_1 is attached by a spring of stiffness K and unstretched length L to another particle of mass m_2. A vertical gravitational force also acts on each particle. The mass m_1 is free to move on a smooth horizontal rail, and the second particle is free to move on a vertical plane. We wish to show that the linear momentum of the system in the $\mathbf{E}_x$ direction is conserved, and then examine what this means for the motion of the system.

---

[2] This can, perhaps, be appreciated by considering the examples discussed below.

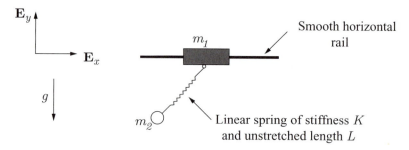

FIGURE 7.2. The cart and the pendulum

Kinematics

We first give some details on the kinematics of the system by defining the position vectors of both particles:

$$\mathbf{r}_1 = x\mathbf{E}_x + y_0\mathbf{E}_y + z_0\mathbf{E}_z, \quad \mathbf{r}_2 = \mathbf{r}_1 + r\mathbf{e}_r,$$

where y_0 and z_0 are constants and the position vector of m_2 relative to m_1 is described using a cylindrical polar coordinate system. As expected, the position vector of the center of mass of the system lies at some point along the spring:

$$\mathbf{r} = \frac{1}{m_1 + m_2}(m_1\mathbf{r}_1 + m_2\mathbf{r}_2) = \mathbf{r}_1 + \frac{m_2}{m_1 + m_2}r\mathbf{e}_r.$$

We can differentiate these position vectors in the usual manner to obtain the velocities and accelerations of the mass particles and the center of mass. Here, we record only the velocity vectors:

$$\mathbf{v}_1 = \dot{x}\mathbf{E}_x, \quad \mathbf{v}_2 = \dot{x}\mathbf{E}_x + \dot{r}\mathbf{e}_r + r\dot{\theta}\mathbf{e}_\theta,$$

$$\mathbf{v} = \dot{x}\mathbf{E}_x + \frac{m_2}{m_1 + m_2}(\dot{r}\mathbf{e}_r + r\dot{\theta}\mathbf{e}_\theta).$$

The acceleration vectors can be obtained in the usual manner. They are used below.

Forces

We now turn to the free-body diagrams for the individual particles and the system of particles. These are shown in Figure 7.3. The spring forces are (see Chapter 4)

$$\mathbf{F}_{s1} = -\mathbf{F}_{s2} = -K(\|\mathbf{r}_1 - \mathbf{r}_2\| - L)\frac{\mathbf{r}_1 - \mathbf{r}_2}{\|\mathbf{r}_1 - \mathbf{r}_2\|} = K(r - L)\mathbf{e}_r.$$

You should note that these forces do not appear in the free-body diagram of the system. The normal forces acting on the particles are

$$\mathbf{N}_1 = N_{1y}\mathbf{E}_y + N_{1z}\mathbf{E}_z, \quad \mathbf{N}_2 = N_{2z}\mathbf{E}_z.$$

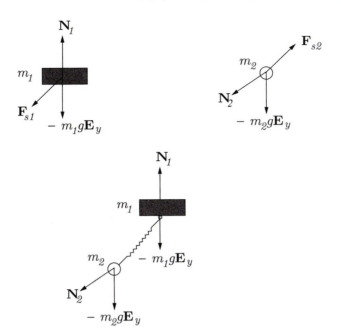

FIGURE 7.3. Free-body diagrams for the cart, pendulum, and system

Balance Laws

We now consider the balances of linear momentum for the individual particles. For the particle of mass m_1 and the particle of mass m_2, we have, respectively,

$$-m_1 g \mathbf{E}_y + N_{1y} \mathbf{E}_y + N_{1z} \mathbf{E}_z + K(r - L)\mathbf{e}_r = m_1 \ddot{x} \mathbf{E}_x \,,$$

$$-m_2 g \mathbf{E}_y + N_{2z} \mathbf{E}_z - K(r - L)\mathbf{e}_r = m_2 \ddot{x} \mathbf{E}_x + m_2 (\ddot{r} - r\dot{\theta}^2)\mathbf{e}_r + m_2 (r\ddot{\theta} + 2\dot{r}\dot{\theta})\mathbf{e}_\theta \,.$$

These are 6 equations to determine the 3 unknowns $\mathbf{N}_1$ and $\mathbf{N}_2$ and provide differential equations for r, x, and θ as functions of time.

With very little work, we find that

$$\mathbf{N}_1 = (m_1 g + K(r - L)\sin(\theta))\mathbf{E}_y \,, \quad \mathbf{N}_2 = \mathbf{0} \,.$$

Next, we obtain a set of (nonlinear) coupled second-order ordinary differential equations:

$$
\begin{aligned}
m_1 \ddot{x} &= K(r - L)\cos(\theta) \,, \\
m_2 \ddot{x} \cos(\theta) + m_2 (\ddot{r} - r\dot{\theta}^2) &= -K(r - L) - m_2 g \sin(\theta) \,, \\
-m_2 \ddot{x} \sin(\theta) + m_2 (r\ddot{\theta} + 2\dot{r}\dot{\theta}) &= -m_2 g \cos(\theta) \,.
\end{aligned}
$$

Notice how the motions of the two particles are coupled. Given a set of initial conditions, $r(t_0)$, $\theta(t_0)$, $x(t_0)$, $\dot{r}(t_0)$, $\dot{\theta}(t_0)$, and $\dot{x}(t_0)$, the solution,

$r(t)$, $\theta(t)$, and $x(t)$, of these equations gives the motions $\mathbf{r}_1(t)$ and $\mathbf{r}_2(t)$ of the particles. Such an analysis is beyond the scope of an undergraduate engineering dynamics course.

Analysis

Next, we consider the balance of linear momentum for the system of particles:[3]

$$-(m_2 + m_1)g\mathbf{E}_y + \mathbf{N}_1 + \mathbf{N}_2 \;=\; (m_1 + m_2)\ddot{x}\mathbf{E}_x + m_2(\ddot{r} - r\dot{\theta}^2)\mathbf{e}_r$$
$$+ m_2(r\ddot{\theta} + 2\dot{r}\dot{\theta})\mathbf{e}_\theta \,.$$

We see immediately that $\mathbf{F}\cdot\mathbf{E}_x = \mathbf{0}$.[4] In other words, the linear momentum of the system in the $\mathbf{E}_x$ direction is conserved. This momentum is

$$\begin{aligned}
\mathbf{G}\cdot\mathbf{E}_x &= (m_1 + m_2)\mathbf{v}\cdot\mathbf{E}_x \\
&= (m_1 + m_2)\dot{x} + m_2(\dot{r}\cos(\theta) - r\dot{\theta}\sin(\theta))\,.
\end{aligned}$$

As a result of this conservation, the velocity of the center of mass in the $\mathbf{E}_x$ direction is constant, and the masses m_1 and m_2 move in such a manner as to preserve this constant velocity. If the initial value of $\mathbf{G}\cdot\mathbf{E}_x$ is equal to G_0, then the motion of the particle of mass m_1 is such that

$$\dot{x} = \frac{1}{m_1 + m_2}(G_0 - m_2(\dot{r}\cos(\theta) - r\dot{\theta}\sin(\theta)))\,.$$

We shall return to this example later on to show that the total energy of this system is conserved.[5]

7.4   Conservation of Angular Momentum

In Section 2 we noted that the angular momentum of the system of particles relative to an arbitrary point P is

$$\mathbf{H}_P = \mathbf{H}_C + (\mathbf{r} - \mathbf{r}_P) \times \mathbf{G}\,,$$

where

$$\mathbf{H}_C = \sum_{k=1}^{n}(\mathbf{r}_k - \mathbf{r}) \times m_k \mathbf{v}_k$$

---

[3]Due to the equal and opposite nature of the spring forces acting on the particles, this equation is equivalent to the addition of the balances of linear momentum for the individual particles. Because the spring is assumed to be massless, the equal and opposite nature of the aforementioned forces can be interpreted using Newton's third law.

[4]That is, $\mathbf{c} = \mathbf{E}_x$.

[5]It is a good exercise to convince yourself that a similar result for linear momentum conservation applies when one replaces the spring with a rigid rod of length L.

is the angular momentum of the system of particles relative to its center of mass C.

We first calculate an expression for its rate of change with respect to time:

$$\dot{\mathbf{H}}_P = \sum_{k=1}^{n} ((\mathbf{v}_k - \mathbf{v}) \times m_k \mathbf{v}_k + (\mathbf{r}_k - \mathbf{r}) \times m_k \mathbf{a}_k)$$
$$+ (\mathbf{v} - \mathbf{v}_P) \times \mathbf{G} + (\mathbf{r} - \mathbf{r}_P) \times \dot{\mathbf{G}},$$

where we have used the product rule. Invoking the balances of linear momentum, $\mathbf{F}_i = m_i \mathbf{a}_i$ and $\mathbf{F} = m\mathbf{a} = \dot{\mathbf{G}}$, the above equation becomes

$$\dot{\mathbf{H}}_P = \sum_{k=1}^{n} (\mathbf{v}_k - \mathbf{v}) \times m_k \mathbf{v}_k + (\mathbf{r}_k - \mathbf{r}) \times \mathbf{F}_k$$
$$+ (\mathbf{v} - \mathbf{v}_P) \times \mathbf{G} + (\mathbf{r} - \mathbf{r}_P) \times \mathbf{F}.$$

We next eliminate those terms that are zero on the right-hand side of this equation with the partial assistance of the identities

$$\mathbf{v} \times \mathbf{G} = \mathbf{0}, \quad \sum_{k=1}^{n} -\mathbf{v} \times m_k \mathbf{v}_k = -\mathbf{v} \times m\mathbf{v} = \mathbf{0}.$$

The final result now follows:

$$\dot{\mathbf{H}}_P = \sum_{k=1}^{n} (\mathbf{r}_k - \mathbf{r}_P) \times \mathbf{F}_k - \mathbf{v}_P \times \mathbf{G}.$$

This result is known as the angular momentum theorem for a system of particles.

There are two important special cases of this result. First, when P is a fixed point O, where $\mathbf{r}_O = \mathbf{0}$, then

$$\dot{\mathbf{H}}_O = \sum_{k=1}^{n} \mathbf{r}_k \times \mathbf{F}_k.$$

Second, when C is the center of mass,

$$\dot{\mathbf{H}}_C = \sum_{k=1}^{n} (\mathbf{r}_k - \mathbf{r}) \times \mathbf{F}_k.$$

For both of these cases we have the usual interpretation that the rate of change of angular momentum relative to a point is the resultant moment due to the forces acting on each particle. However, for the arbitrary case where P may be moving this interpretation does not hold.

We next consider the conditions whereby a component of the angular momentum $\mathbf{H}_P$ in the direction of a given vector $\mathbf{c} = \mathbf{c}(t)$ is conserved. The result parallels that for a single particle. However, here we allow for the possibility that P is moving. For a given vector $\mathbf{c}$, which may be a function of time, we wish to determine when

$$\frac{d}{dt}(\mathbf{H}_P \cdot \mathbf{c}) = 0.$$

Using the previous results, we find that for this conservation it is necessary, and sufficient, that

$$\left(\sum_{k=1}^{n} (\mathbf{r}_k - \mathbf{r}_P) \times \mathbf{F}_k - \mathbf{v}_P \times \mathbf{G} \right) \cdot \mathbf{c} + \mathbf{H}_P \cdot \dot{\mathbf{c}} = 0.$$

For a given problem and a specific point P, it is very difficult to find $\mathbf{c}$ such that this equation holds. Indeed, in most posed problems, P is either the center of mass C or an origin O, while $\mathbf{c} = \mathbf{E}_z$.

7.4.1   Four Whirling Particles

The main class of problems where angular momentum conservation is useful is the mechanism shown in Figure 7.4. Here, four particles are attached to a vertical axle, by springs of stiffness K_i and unstretched length L_i ($i = 1$, 2, 3, or 4). The particles are free to move on smooth horizontal rails. The rails and axle are free to rotate about the vertical with an angular speed ω. We now examine why $\mathbf{H}_O \cdot \mathbf{E}_z$ is conserved in this problem.

Kinematics

For each of the four particles, one defines a cylindrical polar coordinate system $\{r_i, \theta_i, z_i\}$. In particular, $d\theta_i/dt = \omega$. We then have the usual results:

$$\mathbf{r}_i = r_i \mathbf{e}_{ri} + z_0 \mathbf{E}_z, \quad \mathbf{v}_i = \dot{r}_i \mathbf{e}_{ri} + r_i \omega \mathbf{e}_{\theta i},$$

where $z_0 = 0$. The angular momentum of the system relative to the fixed point O is easily calculated using the definition of this angular momentum to be

$$\mathbf{H}_O = (m_1 r_1^2 + m_2 r_2^2 + m_3 r_3^2 + m_4 r_4^2)\omega \mathbf{E}_z.$$

Forces

Leaving the free-body diagrams as an exercise, the resultant force on each particle is

$$\mathbf{F}_i = -K_i(r_i - L_i)\mathbf{e}_{ri} + (N_{i\theta})\mathbf{e}_{\theta i} + (N_{iz} - m_i g)\mathbf{E}_z,$$

where $i = 1$, 2, 3, or 4.

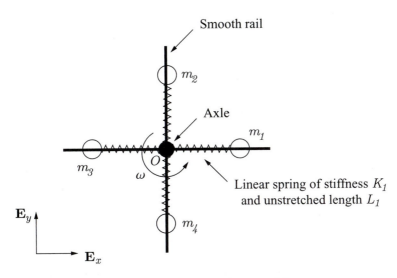

FIGURE 7.4. Four whirling particles

Balance Laws

From $\mathbf{F}_i = m_i \mathbf{a}_i$, one finds that $N_{iz} = m_i g$, as expected. Performing a balance of angular momentum relative to O of the rails and axis in the $\mathbf{E}_z$ direction, we find, on ignoring the inertias of the rails and axis, that if there is no applied moment in the $\mathbf{E}_z$ direction, then[6]

$$r_1 N_{1\theta} + r_2 N_{2\theta} + r_3 N_{3\theta} + r_4 N_{4\theta} = 0.$$

Analysis

We are now in a position to show that $\mathbf{H}_O \cdot \mathbf{E}_z$ is conserved. A direct calculation shows that this is the case:

$$\frac{d}{dt}(\mathbf{H}_0 \cdot \mathbf{E}_z) = \sum_{k=1}^{4}(\mathbf{r}_k \times \mathbf{F}_k) \cdot \mathbf{E}_z$$

$$= \sum_{k=1}^{4}(\mathbf{r}_k \times (-K_k(r_k - L_k)\mathbf{e}_{rk} + N_{k\theta}\mathbf{e}_{\theta k})) \cdot \mathbf{E}_z$$

$$= r_1 N_{1\theta} + r_2 N_{2\theta} + r_3 N_{3\theta} + r_4 N_{4\theta} = 0.$$

It follows that

$$\mathbf{H}_0 \cdot \mathbf{E}_z = (m_1 r_1^2 + m_2 r_2^2 + m_3 r_3^2 + m_4 r_4^2)\omega$$

---

[6]Strictly speaking, the axle and rails constitutes a rigid body, so this result will become clearer to you when we deal with kinematics of rigid bodies later on.

is constant. Here, if one knows the location of the particles and the speed ω at one instant, then, for example, given the location of the particles at a later time one can determine ω.

7.5   Work, Energy, and Conservative Forces

We first recall that for each particle in a system of n particles, we have the work-energy theorem: $\dot{T}_i = \mathbf{F}_i \cdot \mathbf{v}_i$. Recalling that the kinetic energy T of the system of particles is the sum of the individual kinetic energies, one immediately has the work-energy theorem for the system of particles:

$$\frac{dT}{dt} = \sum_{k=1}^{n} \mathbf{F}_k \cdot \mathbf{v}_k \,.$$

We could start here and, paralleling the developments of Chapter 5, develop a theorem for the conservation of the total energy of the system of particles. However, we do not pursue such results here and instead examine the two examples previously discussed and show how the total energy is conserved. We shall then make some general comments at the conclusion of this section.

7.5.1   The Cart and the Pendulum

We first consider the system of two particles discussed in Section 3. For this system, the work-energy theorem gives

$$\begin{aligned}
\frac{dT}{dt} &= \frac{d}{dt}\left(\frac{1}{2}m_1\mathbf{v}_1 \cdot \mathbf{v}_1 + \frac{1}{2}m_2\mathbf{v}_2 \cdot \mathbf{v}_2\right) \\
&= (\mathbf{F}_{s1} - m_1 g\mathbf{E}_y + \mathbf{N}_1) \cdot \mathbf{v}_1 + (\mathbf{F}_{s2} - m_2 g\mathbf{E}_y + \mathbf{N}_2) \cdot \mathbf{v}_2 \,.
\end{aligned}$$

Now, the normal forces are perpendicular to the velocities: $\mathbf{N}_1 \cdot \mathbf{v}_1 = \mathbf{N}_2 \cdot \mathbf{v}_2 = 0$. Secondly,[7]

$$\begin{aligned}
\mathbf{F}_{s1} \cdot \mathbf{v}_1 + \mathbf{F}_{s2} \cdot \mathbf{v}_2 &= -K(\|\mathbf{r}_1 - \mathbf{r}_2\| - L)\left(\frac{\mathbf{r}_1 - \mathbf{r}_2}{\|\mathbf{r}_1 - \mathbf{r}_2\|}\right) \cdot (\mathbf{v}_1 - \mathbf{v}_2) \\
&= -\frac{d}{dt}\left(\frac{K}{2}(\|\mathbf{r}_1 - \mathbf{r}_2\| - L)^2\right) .
\end{aligned}$$

In summary,

$$\frac{dT}{dt} = -\frac{d}{dt}\left(\frac{K}{2}(\|\mathbf{r}_1 - \mathbf{r}_2\| - L)^2 + m_1 g\mathbf{E}_y \cdot \mathbf{v}_1 + m_2 g\mathbf{E}_y \cdot \mathbf{v}_2\right) .$$

---

[7]To establish this result, one uses results from Section 4 of Chapter 5, replacing $\mathbf{r}_D$ with $\mathbf{r}_2$ and noting that $\mathbf{v}_2 \neq \mathbf{0}$.

It follows that the total energy of the system of particles is conserved:

$$\frac{d}{dt}\left(E = T + \frac{K}{2}(\|\mathbf{r}_1 - \mathbf{r}_2\| - L)^2 + m_1 g \mathbf{E}_y \cdot \mathbf{r}_1 + m_2 g \mathbf{E}_y \cdot \mathbf{r}_2\right) = 0.$$

In this example we could replace the spring by a rigid massless rod of length L. The total energy of this system will again be conserved. We now show this. For this system, one has the kinematical results

$$\mathbf{r}_2 - \mathbf{r}_1 = L\mathbf{e}_r, \quad \mathbf{v}_2 - \mathbf{v}_1 = L\dot{\theta}\mathbf{e}_\theta, \quad (\mathbf{r}_2 - \mathbf{r}_1) \cdot (\mathbf{v}_2 - \mathbf{v}_1) = 0.$$

Starting from the work-energy theorem

$$\frac{dT}{dt} = (S\mathbf{e}_r - m_1 g \mathbf{E}_y + \mathbf{N}_1) \cdot \mathbf{v}_1 + (-S\mathbf{e}_r - m_2 g \mathbf{E}_y + \mathbf{N}_2) \cdot \mathbf{v}_2.$$

Here, $S\mathbf{e}_r$ is the tension force in the rod. Again, the normal forces are perpendicular to the velocity vectors, and with the help of the kinematical results above, we conclude energy conservation:

$$\begin{aligned}
\frac{dE}{dt} &= \frac{d}{dt}(T + m_1 g \mathbf{E}_y \cdot \mathbf{r}_1 + m_2 g \mathbf{E}_y \cdot \mathbf{r}_2) \\
&= S\mathbf{e}_r \cdot (\mathbf{v}_1 - \mathbf{v}_2) = S\mathbf{e}_r \cdot (L\dot{\theta}\mathbf{e}_\theta) = 0.
\end{aligned}$$

Notice that the tension force does work on each of the particles. However, its combined power is zero.

7.5.2 Four Whirling Particles

Our final example is the system discussed in Section 4. Here, the work-energy theorem is

$$\begin{aligned}
\frac{dT}{dt} &= \sum_{i=1}^{4} \mathbf{F}_i \cdot \mathbf{v}_i \\
&= \sum_{i=1}^{4} (-K_i(r_i - L_i)\mathbf{e}_{ri} + (N_{i\theta})\mathbf{e}_{\theta i} + (N_{iz} - m_i g)\mathbf{E}_z) \cdot \mathbf{v}_i.
\end{aligned}$$

Simplifying the right-hand side of this equation, we obtain the result

$$\frac{dT}{dt} = \sum_{i=1}^{4} (-K_i(r_i - L_i)\dot{r}_i + (r_i N_{i\theta})\omega).$$

We note that

$$K_i(r_i - L_i)\dot{r}_i = \frac{d}{dt}\left(\frac{K_i}{2}(r_i - L_i)^2\right).$$

That is,

$$K_i(\|r_i\| - L_i)\mathbf{e}_{ri} \cdot \mathbf{v}_i = \frac{d}{dt}\left(\frac{K_i}{2}(\|\mathbf{r}_i\| - L_i)^2\right).$$

Further, after recalling the result that

$$r_1 N_{1\theta} + r_2 N_{2\theta} + r_3 N_{3\theta} + r_4 N_{4\theta} = 0,$$

we find that

$$\frac{d}{dt}\left(E = \frac{1}{2}\sum_{i=1}^{4} m_i \mathbf{v}_i \cdot \mathbf{v}_i + K_i(\|\mathbf{r}_i\| - L_i)^2\right) = 0.$$

In other words, the total energy of the system is conserved.

7.5.3   Comment

In problems involving systems of particles one uses energy conservation in an identical manner as in dealing with a single particle. A subtle feature of systems of particles is that the energy of the individual particles may not be conserved, but the sum of their energies is. This feature is present in the examples discussed above.

7.6   Summary

This chapter was devoted to the kinematics and kinetics of a system of n particles. The first new concept that was introduced was the center of mass C of the system of particles:

$$\mathbf{r} = \frac{1}{m}\sum_{k=1}^{n} m_k \mathbf{r}_k,$$

where $m = \sum_{k=1}^{n} m_k$ is the total mass of the system of particles. We then described how the linear momentum $\mathbf{G}$ of the system of particles was equal to the sum of their linear momenta:

$$\mathbf{G} = m\mathbf{v} = m\dot{\mathbf{r}} = \frac{1}{m}\sum_{k=1}^{n} m_k \mathbf{v}_k.$$

Following this, the angular momentum $\mathbf{H}_P$ of the system of particles was shown to be the angular momentum of the center of mass plus the angular momentum $\mathbf{H}_C$ of the system of particles relative to their center of mass:

$$\mathbf{H}_P = \sum_{k=1}^{n} (\mathbf{r}_k - \mathbf{r}_P) \times m_k \mathbf{v}_k = (\mathbf{r} - \mathbf{r}_P) \times m\mathbf{v} + \mathbf{H}_C,$$

where

$$\mathbf{H}_C = \sum_{k=1}^{n} (\mathbf{r}_k - \mathbf{r}) \times m_k \mathbf{v}_k .$$

Finally, the kinetic energy T of the system of particle was defined to be the sum of the kinetic energies of the individual particles.

Once the kinematical quantities for a system of particles were defined, their kinetics were discussed. By combining the balances of linear momentum for each particle, it was shown that

$$\mathbf{F} = \sum_{k=1}^{n} \mathbf{F}_k = \dot{\mathbf{G}} = m\dot{\mathbf{v}} .$$

This equation was used in Section 3 to establish a linear momentum conservation result. The following angular momentum theorems were established in Section 4:

$$\dot{\mathbf{H}}_O = \sum_{k=1}^{n} \mathbf{r}_k \times \mathbf{F}_k , \quad \dot{\mathbf{H}}_C = \sum_{k=1}^{n} (\mathbf{r}_k - \mathbf{r}) \times \mathbf{F}_k ,$$

where O is a fixed point. These results were then used to show when a component of an angular momentum was conserved. In Section 5, the work-energy theorem for a system of particles was discussed:

$$\dot{T} = \sum_{k=1}^{n} \mathbf{F}_k \cdot \mathbf{v}_k .$$

This theorem was the starting point for proving energy conservation for systems of particles.

7.7 Exercises

The following short exercises are intended to assist you in reviewing Chapter 7.

7.1 Starting from the definition of the position vector of the center of mass, show that

$$\sum_{k=1}^{n} m_k (\mathbf{r}_k - \mathbf{r}) = \mathbf{0}, \quad \sum_{k=1}^{n} m_k (\mathbf{v}_k - \mathbf{v}) = \mathbf{0} .$$

Where were these identities used?

7.2 Starting from the definition of the angular momentum of a system of particles relative to a point P, prove that

$$\mathbf{H}_P = (\mathbf{r} - \mathbf{r}_P) \times m\mathbf{v} + \mathbf{H}_C .$$

7.3 Starting from the definition of the kinetic energy T of a system of particles, show that

$$T = \frac{1}{2}m\mathbf{v} \cdot \mathbf{v} + \frac{1}{2}\sum_{k=1}^{n} m_k \left(\mathbf{v}_k - \mathbf{v}\right) \cdot \left(\mathbf{v}_k - \mathbf{v}\right).$$

Using this result, show that the kinetic energy of a system of particles is not, in general, equal to the kinetic energy of the center of mass.

7.4 Consider two particles which are free to move on a horizontal surface $z = 0$. Vertical gravitational forces $-m_1 g \mathbf{E}_z$ and $-m_2 g \mathbf{E}_z$ act on the respective particles. The position vectors of the particles are $\mathbf{r}_1 = x\mathbf{E}_x + y\mathbf{E}_y$ and $\mathbf{r}_2 = \mathbf{r}_1 + r\mathbf{e}_r$. Derive an expression for the position vector $\mathbf{r}$ of the center of mass C of this system of particles. Verify your answer by examining the limiting cases that m_1 is much larger than m_2 and vice versa.

7.5 Consider the system of particles discussed in Exercise 7.4. Suppose the particles are connected by a linear spring of stiffness K and unstretched length L. Show that the linear momenta $\mathbf{G} \cdot \mathbf{E}_x$ and $\mathbf{G} \cdot \mathbf{E}_y$ are conserved. What do these results imply about the motion of the center of mass C of this system of particles?

7.6 For the system of particles discussed in Exercise 7.5, prove that $\mathbf{H}_C \cdot \mathbf{E}_z$ is conserved. What does this result imply about $\dot{\theta}$?

7.7 Consider the system of particles discussed in Exercise 7.5. Starting from the work-energy theorem, prove that the total energy E of the system of particles is conserved. Here,

$$E = \frac{1}{2}\left(m_1\mathbf{v}_1 \cdot \mathbf{v}_1 + m_2\mathbf{v}_2 \cdot \mathbf{v}_2\right) + \frac{K}{2}\left(r - L\right)^2.$$

7.8 For the cart and pendulum system discussed in Section 3.1, show that $\mathbf{G} \cdot \mathbf{E}_x$ and the total energy E are still conserved if the spring is replaced by an inextensible string of length L.

7.9 Consider the system of four particles discussed in Section 4.1. If one had the ability to measure r_1, r_2, r_3, r_4, and $\dot{\theta}$ for this system, how would one verify that $\mathbf{H}_O \cdot \mathbf{E}_z$ was conserved?

8

Planar Kinematics of Rigid Bodies

TOPICS

This chapter contains results on the planar kinematics of rigid bodies. In particular, we show how to establish certain useful representations for the velocity and acceleration vectors of any material point of a rigid body. We also discuss the angular velocity vector of a rigid body. These concepts are illustrated using two important applications: mechanisms and rolling rigid bodies. Finally, we discuss linear $\mathbf{G}$ and angular $(\mathbf{H}, \mathbf{H}_O, \mathbf{H}_A)$ momenta of rigid bodies and the inertias that accompany them.[1]

8.1   The Motion of a Rigid Body

8.1.1   General Considerations

A body $\mathcal{B}$ is a collection of material points (mass particles or particles). We denote a material point by X. The position of the material point X, relative to a fixed origin, at time t is denoted by $\mathbf{x}$ (see Figure 8.1). The present (or current) configuration $\boldsymbol{\kappa}_t$ of the body is a smooth, one-to-one,

---

[1] The details presented here are far more advanced than those in most undergraduate texts. This is partially because the presentation is influenced by the recent renaissance in continuum mechanics. We mention in particular the influential works by Beatty [5] and Casey [11, 12, 13, 14], who used the fruits of this era to present enlightening treatments of rigid body mechanics. This chapter is based on the aforementioned works and Chapter 4 of Gurtin [30].

onto function that has a continuous inverse. It maps material points X of $\mathcal{B}$ to points in three-dimensional Euclidean space: $\mathbf{x} = \kappa_t(X)$. As the location $\mathbf{x}$ of the particle X changes with time, this function depends on time, hence the subscript t.

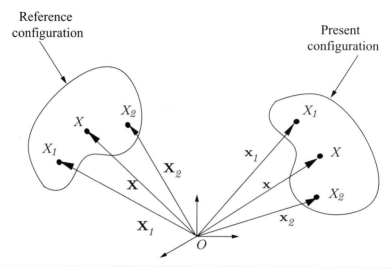

FIGURE 8.1. Configurations of a body $\mathcal{B}$

It is convenient to define a fixed reference configuration κ_0 of the body. This configuration is defined by the function $\mathbf{X} = \kappa_0(X)$. Since this function is invertible, we can use the position vector $\mathbf{X}$ of a material point X in the reference configuration to uniquely define the material point of interest. One can then define the motion of the body as a function of $\mathbf{X}$ and t:

$$\mathbf{x} = \chi(\mathbf{X}, t).$$

Notice that the motion of a material point of $\mathcal{B}$ depends on time and the material point of interest. To see this imagine a compact disc spinning in a CD player. The motion of a particle on the outer rim of the disc is clearly different from the motion of a particle on the inner rim of the disc. Further, the place in space that each of these particles occupies depends on the time t of interest.

The previous developments are general and are used in continuum mechanics, a field which encompasses the mechanics of solids and fluids.

8.1.2 A Rigid Body

For rigid bodies, the nature of the function $\chi(\mathbf{X}, t)$ can be simplified dramatically. We refer to the rigid motion as $\mathbf{x} = \chi_R(\mathbf{X}, t)$. First, for rigid bodies the distance between *any* two mass particles, say X_1 and X_2, remains constant for all motions. Mathematically, this is equivalent to saying that

$$\|\mathbf{x}_1 - \mathbf{x}_2\| = \|\mathbf{X}_1 - \mathbf{X}_2\|.$$

Secondly, the motion of the rigid body preserves orientations. Using a classical result,[2] it can be proven that the motion of a rigid body has the form

$$\left[\begin{array}{c} (\mathbf{x}_1 - \mathbf{x}_2) \cdot \mathbf{E}_x \\ (\mathbf{x}_1 - \mathbf{x}_2) \cdot \mathbf{E}_y \\ (\mathbf{x}_1 - \mathbf{x}_2) \cdot \mathbf{E}_z \end{array} \right] = \left[\begin{array}{ccc} Q_{11}(t) & Q_{12}(t) & Q_{13}(t) \\ Q_{21}(t) & Q_{22}(t) & Q_{23}(t) \\ Q_{31}(t) & Q_{32}(t) & Q_{33}(t) \end{array} \right] \left[\begin{array}{c} (\mathbf{X}_1 - \mathbf{X}_2) \cdot \mathbf{E}_x \\ (\mathbf{X}_1 - \mathbf{X}_2) \cdot \mathbf{E}_y \\ (\mathbf{X}_1 - \mathbf{X}_2) \cdot \mathbf{E}_z \end{array} \right].$$

Here, the matrix whose components are $Q_{11}, \ldots, Q_{33}$ is a proper-orthogonal or rotation matrix.

To abbreviate the subsequent developments, we define

$$\mathbf{Q} = \left[\begin{array}{ccc} Q_{11}(t) & Q_{12}(t) & Q_{13}(t) \\ Q_{21}(t) & Q_{22}(t) & Q_{23}(t) \\ Q_{31}(t) & Q_{32}(t) & Q_{33}(t) \end{array} \right],$$

$$\mathbf{Q}^T = \left[\begin{array}{ccc} Q_{11}(t) & Q_{21}(t) & Q_{31}(t) \\ Q_{12}(t) & Q_{22}(t) & Q_{32}(t) \\ Q_{13}(t) & Q_{23}(t) & Q_{33}(t) \end{array} \right],$$

$$\mathbf{I} = \left[\begin{array}{ccc} 1 & 0 & 0 \\ 0 & 1 & 0 \\ 0 & 0 & 1 \end{array} \right].$$

The relations satisfied by any rotation matrix $\mathbf{Q}$ can be written as follows:

$$\mathbf{Q}\mathbf{Q}^T = \mathbf{I}, \quad \det(\mathbf{Q}) = 1.$$

In words, the inverse of a rotation matrix is its transpose, and the determinant of a rotation matrix is 1. The relations $\mathbf{Q}\mathbf{Q}^T = \mathbf{I}$ can be interpreted as 6 conditions on the 9 components of $\mathbf{Q}$. It follows that only 3 of the 9 components Q_{ik} are independent. One then has the problem of parametrizing $\mathbf{Q}$ in terms of 3 independent parameters. There are several methods of doing this: Euler angles and Euler parameters being the most popular.[3] In an undergraduate engineering dynamics course one considers a particular one-parameter family of rotation matrices. Finally, we note that $\det(\mathbf{Q}) = 1$ implies that $\mathbf{Q}$ preserves orientations.

---

[2] The proof of this result is beyond the scope of this course. One proof may be found on pages 49–50 of Gurtin [30]. A good discussion on the relationship between this result with Euler's theorem on the motion of a rigid body and Chasles' theorem can be found in Beatty [5] (see also Beatty [4]). Euler's representation of rigid body motion can be seen on pages 30–32 of Euler [24].

[3] Details on these parametrizations can be found, for example, in Beatty [5], Casey [12, 14], Greenwood [29], Shuster [58], Synge and Griffith [64], and Whittaker [67].

Because $\mathbf{Q}$ is a rotation matrix, we note that[4]

$$
\begin{aligned}
0 &= \frac{d\mathbf{I}}{dt} = \frac{d(\mathbf{Q}\mathbf{Q}^T)}{dt} \\
&= \frac{d\mathbf{Q}}{dt}\mathbf{Q}^T + \mathbf{Q}\frac{d\mathbf{Q}^T}{dt} .
\end{aligned}
$$

That is,

$$
\frac{d\mathbf{Q}}{dt}\mathbf{Q}^T = -\mathbf{Q}\frac{d\mathbf{Q}^T}{dt} = -\left(\frac{d\mathbf{Q}}{dt}\mathbf{Q}^T\right)^T .
$$

Hence, $\mathbf{Q}\dot{\mathbf{Q}}^T$ is a skew-symmetric matrix:

$$
\dot{\mathbf{Q}}\mathbf{Q}^T =
\begin{bmatrix}
0 & -\Omega_{21} & \Omega_{13} \\
\Omega_{21} & 0 & -\Omega_{32} \\
-\Omega_{13} & \Omega_{32} & 0
\end{bmatrix} .
$$

The three components Ω_{21}, Ω_{13}, and Ω_{32} can be expressed in terms of the components of $\mathbf{Q}$ and its time derivatives, but we leave this as an exercise.

8.1.3 Angular Velocity and Acceleration Vectors

Returning to the discussion of a rigid body, we recall that

$$
\begin{bmatrix}
(\mathbf{x}_1 - \mathbf{x}_2) \cdot \mathbf{E}_x \\
(\mathbf{x}_1 - \mathbf{x}_2) \cdot \mathbf{E}_y \\
(\mathbf{x}_1 - \mathbf{x}_2) \cdot \mathbf{E}_z
\end{bmatrix}
=
\begin{bmatrix}
Q_{11}(t) & Q_{12}(t) & Q_{13}(t) \\
Q_{21}(t) & Q_{22}(t) & Q_{23}(t) \\
Q_{31}(t) & Q_{32}(t) & Q_{33}(t)
\end{bmatrix}
\begin{bmatrix}
(\mathbf{X}_1 - \mathbf{X}_2) \cdot \mathbf{E}_x \\
(\mathbf{X}_1 - \mathbf{X}_2) \cdot \mathbf{E}_y \\
(\mathbf{X}_1 - \mathbf{X}_2) \cdot \mathbf{E}_z
\end{bmatrix} .
$$

Since the matrix $\mathbf{Q}$ is a rotation matrix, we can easily invert this relationship by multiplying both sides by the transpose of $\mathbf{Q}$:

$$
\begin{bmatrix}
(\mathbf{X}_1 - \mathbf{X}_2) \cdot \mathbf{E}_x \\
(\mathbf{X}_1 - \mathbf{X}_2) \cdot \mathbf{E}_y \\
(\mathbf{X}_1 - \mathbf{X}_2) \cdot \mathbf{E}_z
\end{bmatrix}
=
\begin{bmatrix}
Q_{11}(t) & Q_{21}(t) & Q_{31}(t) \\
Q_{12}(t) & Q_{22}(t) & Q_{32}(t) \\
Q_{13}(t) & Q_{23}(t) & Q_{33}(t)
\end{bmatrix}
\begin{bmatrix}
(\mathbf{x}_1 - \mathbf{x}_2) \cdot \mathbf{E}_x \\
(\mathbf{x}_1 - \mathbf{x}_2) \cdot \mathbf{E}_y \\
(\mathbf{x}_1 - \mathbf{x}_2) \cdot \mathbf{E}_z
\end{bmatrix} .
$$

Now let us examine the relationship between the velocity and acceleration vectors of two material points of the body. A simple differentiation, where we note that $\mathbf{X}_1$ and $\mathbf{X}_2$ are constant, gives

$$
\begin{bmatrix}
(\mathbf{v}_1 - \mathbf{v}_2) \cdot \mathbf{E}_x \\
(\mathbf{v}_1 - \mathbf{v}_2) \cdot \mathbf{E}_y \\
(\mathbf{v}_1 - \mathbf{v}_2) \cdot \mathbf{E}_z
\end{bmatrix}
=
\begin{bmatrix}
\dot{Q}_{11}(t) & \dot{Q}_{12}(t) & \dot{Q}_{13}(t) \\
\dot{Q}_{21}(t) & \dot{Q}_{22}(t) & \dot{Q}_{23}(t) \\
\dot{Q}_{31}(t) & \dot{Q}_{32}(t) & \dot{Q}_{33}(t)
\end{bmatrix}
\begin{bmatrix}
(\mathbf{X}_1 - \mathbf{X}_2) \cdot \mathbf{E}_x \\
(\mathbf{X}_1 - \mathbf{X}_2) \cdot \mathbf{E}_y \\
(\mathbf{X}_1 - \mathbf{X}_2) \cdot \mathbf{E}_z
\end{bmatrix} .
$$

---

[4]Recall that the transpose of a product of two matrices A and B is $(\mathsf{A}\mathsf{B})^T = \mathsf{B}^T\mathsf{A}^T$. Further, a matrix C is symmetric if $\mathsf{C} = \mathsf{C}^T$, and is skew-symmetric if $\mathsf{C} = -\mathsf{C}^T$.

Here, $\mathbf{v}_1 = \dot{\mathbf{x}}_1$ and $\mathbf{v}_2 = \dot{\mathbf{x}}_2$. We next substitute for $\mathbf{X}_1$ and $\mathbf{X}_2$ and use the earlier observation about $\dot{\mathbf{Q}}\mathbf{Q}^T$ to find that

$$
\begin{bmatrix} (\mathbf{v}_1 - \mathbf{v}_2) \cdot \mathbf{E}_x \\ (\mathbf{v}_1 - \mathbf{v}_2) \cdot \mathbf{E}_y \\ (\mathbf{v}_1 - \mathbf{v}_2) \cdot \mathbf{E}_z \end{bmatrix} = \begin{bmatrix} 0 & -\Omega_{21} & \Omega_{13} \\ \Omega_{21} & 0 & -\Omega_{32} \\ -\Omega_{13} & \Omega_{32} & 0 \end{bmatrix} \begin{bmatrix} (\mathbf{x}_1 - \mathbf{x}_2) \cdot \mathbf{E}_x \\ (\mathbf{x}_1 - \mathbf{x}_2) \cdot \mathbf{E}_y \\ (\mathbf{x}_1 - \mathbf{x}_2) \cdot \mathbf{E}_z \end{bmatrix}.
$$

We can write this important result in vector notation;

$$
\mathbf{v}_1 - \mathbf{v}_2 = \boldsymbol{\omega} \times (\mathbf{x}_1 - \mathbf{x}_2),
$$

where $\boldsymbol{\omega}$ is known as the *angular velocity vector* of the rigid body:

$$
\boldsymbol{\omega} = \Omega_{32}\mathbf{E}_x + \Omega_{13}\mathbf{E}_y + \Omega_{21}\mathbf{E}_z.
$$

You should notice that this vector depends on time and not on the particle of the body: it has the same value for each X. The reason for this is that it is obtained by differentiating $\mathbf{Q}$, and this matrix is a function of t only.

We can easily find the relationships between the accelerations $\mathbf{a}_1$ and $\mathbf{a}_2$ of the material points X_1 and X_2 by differentiating the relationship between their velocities:

$$
\begin{aligned}
\mathbf{a}_1 - \mathbf{a}_2 &= \dot{\mathbf{v}}_1 - \dot{\mathbf{v}}_2 \\
&= \dot{\boldsymbol{\omega}} \times (\mathbf{x}_1 - \mathbf{x}_2) + \boldsymbol{\omega} \times (\mathbf{v}_1 - \mathbf{v}_2) \\
&= \boldsymbol{\alpha} \times (\mathbf{x}_1 - \mathbf{x}_2) + \boldsymbol{\omega} \times (\boldsymbol{\omega} \times (\mathbf{x}_1 - \mathbf{x}_2)).
\end{aligned}
$$

Here, $\boldsymbol{\alpha}$ is the *angular acceleration vector* of the rigid body:

$$
\boldsymbol{\alpha} = \dot{\boldsymbol{\omega}}.
$$

8.1.4   Fixed-Axis Rotation

All of the aforementioned developments are general. In an introductory undergraduate engineering dynamics course one considers a special case. In this special case, the axis of rotation is fixed. This axis is normally taken to coincide with $\mathbf{E}_z$.

For this special case, the rotation matrix $\mathbf{Q}$ has a particularly simple form:

$$
\mathbf{Q} = \begin{bmatrix} \cos(\theta) & -\sin(\theta) & 0 \\ \sin(\theta) & \cos(\theta) & 0 \\ 0 & 0 & 1 \end{bmatrix}.
$$

The angle θ represents a counterclockwise rotation of the rigid body about $\mathbf{E}_z$.

Let us now establish the angular velocity and acceleration vectors associated with this rotation matrix:

$$\dot{Q}Q^T = \dot{\theta} \begin{bmatrix} -\sin(\theta) & -\cos(\theta) & 0 \\ \cos(\theta) & -\sin(\theta) & 0 \\ 0 & 0 & 0 \end{bmatrix} \begin{bmatrix} \cos(\theta) & \sin(\theta) & 0 \\ -\sin(\theta) & \cos(\theta) & 0 \\ 0 & 0 & 1 \end{bmatrix}$$

$$= \dot{\theta} \begin{bmatrix} 0 & -1 & 0 \\ 1 & 0 & 0 \\ 0 & 0 & 0 \end{bmatrix}.$$

Hence, examining the components of the above matrix, we find that

$$\boldsymbol{\omega} = \dot{\theta}\mathbf{E}_z, \quad \boldsymbol{\alpha} = \ddot{\theta}\mathbf{E}_z.$$

8.2   Kinematical Relations and a Corotational Basis

In our previous developments we used a fixed (right-handed) Cartesian basis. It is convenient, when discussing the dynamics of rigid bodies, to introduce another basis which is known as a *corotational* basis.[5] This section discusses such a basis and points out some features of its use.

8.2.1   The Corotational Basis

Here, we define a basis $\{\mathbf{e}_x, \mathbf{e}_y, \mathbf{e}_z\}$ that rotates with the body. As a result, it is known as a corotational basis. Our discussion of this basis follows Casey [11, 14].

Referring to Figure 8.2, we start by picking 4 material points X_1, X_2, X_3, and X_4 of the body. These points are chosen, such that the vectors

$$\mathbf{E}_x = \mathbf{X}_1 - \mathbf{X}_4, \quad \mathbf{E}_y = \mathbf{X}_2 - \mathbf{X}_4, \quad \mathbf{E}_z = \mathbf{X}_3 - \mathbf{X}_4$$

form a fixed, right-handed, Cartesian basis. We next consider the present relative locations of the 4 material points. Because $\mathbf{Q}$ preserves lengths and orientations, the three vectors $\mathbf{x}_1 - \mathbf{x}_4$, $\mathbf{x}_2 - \mathbf{x}_4$, and $\mathbf{x}_3 - \mathbf{x}_4$ will also form a right-handed orthonormal basis.[6] As a result, we define the corotational basis to be

$$\mathbf{e}_x = \mathbf{x}_1 - \mathbf{x}_4, \quad \mathbf{e}_y = \mathbf{x}_2 - \mathbf{x}_4, \quad \mathbf{e}_z = \mathbf{x}_3 - \mathbf{x}_4.$$

---

[5]This basis is often referred to as a body fixed frame or an embedded frame.

[6]The proof of this result is beyond our scope here. A proof may be found in Casey [11]. For the special case of a fixed-axis rotation, we give an explicit demonstration of this result below.

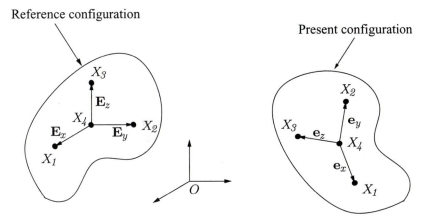

FIGURE 8.2. The corotational basis $\{\mathbf{e}_x, \mathbf{e}_y, \mathbf{e}_z\}$ and the fixed Cartesian basis $\{\mathbf{E}_x, \mathbf{E}_y, \mathbf{E}_z\}$

Since the corotational basis moves with the body, we can use our previous results for relative velocities in Section 1.3,[7] to find that

$$\dot{\mathbf{e}}_x = \boldsymbol{\omega} \times \mathbf{e}_x, \quad \dot{\mathbf{e}}_y = \boldsymbol{\omega} \times \mathbf{e}_y, \quad \dot{\mathbf{e}}_z = \boldsymbol{\omega} \times \mathbf{e}_z.$$

Furthermore, we can differentiate these results to obtain, for example,

$$\ddot{\mathbf{e}}_x = \boldsymbol{\alpha} \times \mathbf{e}_x + \boldsymbol{\omega} \times (\boldsymbol{\omega} \times \mathbf{e}_x).$$

Related results hold for $\ddot{\mathbf{e}}_y$ and $\ddot{\mathbf{e}}_z$. The aforementioned relations will prove useful when we establish certain kinematical results later on.

Since the set $\{\mathbf{e}_x, \mathbf{e}_y, \mathbf{e}_z\}$ is a basis, given any vector $\mathbf{b}$, one has the representations

$$
\begin{aligned}
\mathbf{b} &= b_x \mathbf{e}_x + b_y \mathbf{e}_y + b_z \mathbf{e}_z \\
&= B_x \mathbf{E}_x + B_y \mathbf{E}_y + B_z \mathbf{E}_z.
\end{aligned}
$$

If $\mathbf{b}$ is a constant vector, $\dot{\mathbf{b}} = \mathbf{0}$, then B_x, B_y, and B_z are constant. However, because the corotational basis changes with time, the constancy of $\mathbf{b}$ does not imply that b_x, b_y, and b_z are constant.

8.2.2   The Corotational Basis for the Fixed-Axis Case

As mentioned previously, for the fixed-axis case, the rotation matrix $\mathbf{Q}$ has a particularly simple form:

$$
\mathbf{Q} = \begin{bmatrix} \cos(\theta) & -\sin(\theta) & 0 \\ \sin(\theta) & \cos(\theta) & 0 \\ 0 & 0 & 1 \end{bmatrix}.
$$

---

[7]For example, $\dot{\mathbf{e}}_x = \mathbf{v}_1 - \mathbf{v}_4 = \boldsymbol{\omega} \times (\mathbf{x}_1 - \mathbf{x}_4) = \boldsymbol{\omega} \times \mathbf{e}_x$.

The angle θ represents a counterclockwise rotation of the rigid body about $\mathbf{E}_z$.

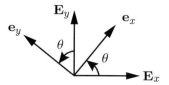

FIGURE 8.3. The corotational basis for a fixed-axis rotation about $\mathbf{E}_z$

Taking the aforementioned four points, we find that

$$
\begin{bmatrix}
(\mathbf{e}_x = \mathbf{x}_1 - \mathbf{x}_4) \cdot \mathbf{E}_x \\
(\mathbf{e}_x = \mathbf{x}_1 - \mathbf{x}_4) \cdot \mathbf{E}_y \\
(\mathbf{e}_x = \mathbf{x}_1 - \mathbf{x}_4) \cdot \mathbf{E}_z
\end{bmatrix}
=
\begin{bmatrix}
\cos(\theta) & -\sin(\theta) & 0 \\
\sin(\theta) & \cos(\theta) & 0 \\
0 & 0 & 1
\end{bmatrix}
\begin{bmatrix}
(\mathbf{E}_x = \mathbf{X}_1 - \mathbf{X}_4) \cdot \mathbf{E}_x \\
(\mathbf{E}_x = \mathbf{X}_1 - \mathbf{X}_4) \cdot \mathbf{E}_y \\
(\mathbf{E}_x = \mathbf{X}_1 - \mathbf{X}_4) \cdot \mathbf{E}_z
\end{bmatrix}
$$

$$
=
\begin{bmatrix}
\cos(\theta) \\
\sin(\theta) \\
0
\end{bmatrix},
$$

$$
\begin{bmatrix}
(\mathbf{e}_y = \mathbf{x}_2 - \mathbf{x}_4) \cdot \mathbf{E}_x \\
(\mathbf{e}_y = \mathbf{x}_2 - \mathbf{x}_4) \cdot \mathbf{E}_y \\
(\mathbf{e}_y = \mathbf{x}_2 - \mathbf{x}_4) \cdot \mathbf{E}_z
\end{bmatrix}
=
\begin{bmatrix}
\cos(\theta) & -\sin(\theta) & 0 \\
\sin(\theta) & \cos(\theta) & 0 \\
0 & 0 & 1
\end{bmatrix}
\begin{bmatrix}
(\mathbf{E}_y = \mathbf{X}_2 - \mathbf{X}_4) \cdot \mathbf{E}_x \\
(\mathbf{E}_y = \mathbf{X}_2 - \mathbf{X}_4) \cdot \mathbf{E}_y \\
(\mathbf{E}_y = \mathbf{X}_2 - \mathbf{X}_4) \cdot \mathbf{E}_z
\end{bmatrix}
$$

$$
=
\begin{bmatrix}
-\sin(\theta) \\
\cos(\theta) \\
0
\end{bmatrix},
$$

$$
\begin{bmatrix}
(\mathbf{e}_z = \mathbf{x}_3 - \mathbf{x}_4) \cdot \mathbf{E}_x \\
(\mathbf{e}_z = \mathbf{x}_3 - \mathbf{x}_4) \cdot \mathbf{E}_y \\
(\mathbf{e}_z = \mathbf{x}_3 - \mathbf{x}_4) \cdot \mathbf{E}_z
\end{bmatrix}
=
\begin{bmatrix}
\cos(\theta) & -\sin(\theta) & 0 \\
\sin(\theta) & \cos(\theta) & 0 \\
0 & 0 & 1
\end{bmatrix}
\begin{bmatrix}
(\mathbf{E}_z = \mathbf{X}_3 - \mathbf{X}_4) \cdot \mathbf{E}_x \\
(\mathbf{E}_z = \mathbf{X}_3 - \mathbf{X}_4) \cdot \mathbf{E}_y \\
(\mathbf{E}_z = \mathbf{X}_3 - \mathbf{X}_4) \cdot \mathbf{E}_z
\end{bmatrix}
$$

$$
=
\begin{bmatrix}
0 \\
0 \\
1
\end{bmatrix}.
$$

In summary, we obtain the results shown graphically in Figure 8.3:

$$
\begin{bmatrix}
\mathbf{e}_x \\
\mathbf{e}_y \\
\mathbf{e}_z
\end{bmatrix}
=
\begin{bmatrix}
\cos(\theta) & \sin(\theta) & 0 \\
-\sin(\theta) & \cos(\theta) & 0 \\
0 & 0 & 1
\end{bmatrix}
\begin{bmatrix}
\mathbf{E}_x \\
\mathbf{E}_y \\
\mathbf{E}_z
\end{bmatrix}.
$$

These relations have a familiar form. It is a useful exercise to verify that the corotational basis is right-handed: $\mathbf{e}_z \cdot (\mathbf{e}_x \times \mathbf{e}_y) = 1$.

For the case of a fixed-axis rotation about $\mathbf{E}_z$, we showed previously that

$$
\boldsymbol{\omega} = \dot{\theta}\mathbf{E}_z, \quad \boldsymbol{\alpha} = \ddot{\theta}\mathbf{E}_z.
$$

We can use these results to find that

$$
\begin{aligned}
\dot{\mathbf{e}}_x &= \boldsymbol{\omega} \times \mathbf{e}_x = \dot{\theta}\mathbf{e}_y, \\
\dot{\mathbf{e}}_y &= \boldsymbol{\omega} \times \mathbf{e}_y = -\dot{\theta}\mathbf{e}_x, \\
\dot{\mathbf{e}}_z &= \boldsymbol{\omega} \times \mathbf{e}_z = \mathbf{0}.
\end{aligned}
$$

Alternatively, we can work directly with the representations for $\mathbf{e}_x$, $\mathbf{e}_y$, and $\mathbf{e}_z$ in terms of $\mathbf{E}_x$, $\mathbf{E}_y$, and $\mathbf{E}_z$ to arrive at the same results.

8.2.3   A Particle Moving on a Rigid Body

It is convenient at this stage to consider an example. As shown in Figure 8.4, a particle moves on the surface of a circular disk. The disk is rotating about the $\mathbf{E}_z$ axis with an angular speed $\dot{\theta} = \omega$ and an angular acceleration $\ddot{\theta} = \alpha$. The center of the disk O is fixed. We seek to determine the velocity vector of the particle and the velocity vector of a point X of the disk.

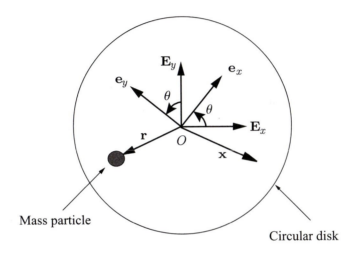

FIGURE 8.4. A particle moving on a rotating disk

For the example of interest, suppose

$$\mathbf{r} = 10t^2 \mathbf{e}_x + 20t\mathbf{e}_y \, .$$

Further, let the position vector of X be

$$\mathbf{x} = x\mathbf{e}_x + y\mathbf{e}_y \, ,$$

where x and y are constants.

To calculate the velocity vectors, we merely differentiate the position vectors and use our previous results on the derivatives of $\mathbf{e}_x$ and $\mathbf{e}_y$:

$$
\begin{aligned}
\dot{\mathbf{r}} &= 20t\mathbf{e}_x + 10t^2\dot{\mathbf{e}}_x + 20\mathbf{e}_y + 20t\dot{\mathbf{e}}_y \\
&= 20t\mathbf{e}_x + 10t^2\omega\mathbf{e}_y + 20\mathbf{e}_y - 20t\omega\mathbf{e}_x \, , \\
\dot{\mathbf{x}} &= x\dot{\mathbf{e}}_x + y\dot{\mathbf{e}}_y \\
&= x\omega\mathbf{e}_y - y\omega\mathbf{e}_x \, .
\end{aligned}
$$

You should notice that

$$\dot{\mathbf{x}} = \boldsymbol{\omega} \times \mathbf{x}, \quad \dot{\mathbf{r}} \neq \boldsymbol{\omega} \times \mathbf{r}.$$

The reason for these results lies in the fact that $\mathbf{x}$ is the position vector of a point of the disk, while $\mathbf{r}$ is the position vector of a particle that moves relative to the disk.

We leave it as an exercise to determine the acceleration vectors of the particle and X.

8.3   Mechanisms

One of the main applications of the theory of rigid bodies is an analysis of the kinematics of mechanisms. Two of the most important mechanisms are the slider crank and the four-bar linkage. In general, elements of mechanisms are deformable bodies, but a primitive analysis assumes that these elements are rigid. Here, we also assume that the motions of these elements are coplanar - but it is not very difficult to consider the more general case.[8]

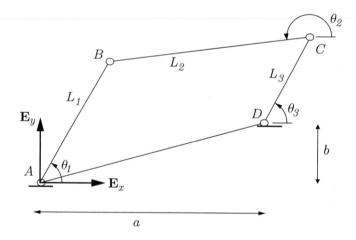

FIGURE 8.5. A four-bar linkage

As an example, consider the four-bar linkage shown in Figure 8.5. Here, the bar AD is fixed:

$$\mathbf{v}_A = \mathbf{v}_D = \mathbf{0}.$$

---

[8]The study of mechanisms is an important area of mechanical engineering. Our discussion here touches but a small part of it. The interested reader is referred to the textbooks of Bottema and Rott [9], Mabie and Ocvirk [38], and Paul [47] for further treatments and issues.

The motion of the bar AB is assumed to be known. In other words, θ_1, $\dot{\theta}_1$, and $\ddot{\theta}_1$ are prescribed. The bars AD, DC, AB, and BC are interconnected by pin-joints. One seeks to determine the motion of the two bars DC and BC. That is, one seeks

$$\theta_2 \,,\, \dot{\theta}_2 \,,\, \ddot{\theta}_2 \,,\, \theta_3 \,,\, \dot{\theta}_3 \,,\, \ddot{\theta}_3 \,,$$

as functions of time. You should note that the angular velocity vectors of the bars AB, BC, and DC are, respectively,

$$\boldsymbol{\omega}_{AB} = \dot{\theta}_1 \mathbf{E}_z \,, \quad \boldsymbol{\omega}_{BC} = \dot{\theta}_2 \mathbf{E}_z \,, \quad \boldsymbol{\omega}_{DC} = \dot{\theta}_3 \mathbf{E}_z \,.$$

First, the linkages are connected together:

$$\mathbf{r}_{DA} = \mathbf{r}_{BA} + \mathbf{r}_{CB} + \mathbf{r}_{DC} \,,$$

where $\mathbf{r}_{DA} = \mathbf{r}_D - \mathbf{r}_A$, etc. Introducing the angles shown in Figure 8.5, we find that this relationship can be written as

$$
\begin{aligned}
a\mathbf{E}_x + b\mathbf{E}_y \;=\; & L_1(\cos(\theta_1)\mathbf{E}_x + \sin(\theta_1)\mathbf{E}_y) \\
& - L_2(\cos(\theta_2)\mathbf{E}_x + \sin(\theta_2)\mathbf{E}_y) - L_3(\cos(\theta_3)\mathbf{E}_x + \sin(\theta_3)\mathbf{E}_y) \,.
\end{aligned}
$$

This constitutes 2 scalar equations for the unknown angles θ_2 and θ_3:

$$
\begin{aligned}
a &= L_1 \cos(\theta_1) - L_2 \cos(\theta_2) - L_3 \cos(\theta_3) \,, \\
b &= L_1 \sin(\theta_1) - L_2 \sin(\theta_2) - L_3 \sin(\theta_3) \,.
\end{aligned}
$$

These equations are nonlinear and, in general, have multiple solutions (θ_2, θ_3). To see this, one merely has to draw different possible configurations of the mechanism.

To obtain a second set of equations, we differentiate the position vector relationship above:

$$\mathbf{v}_{DA} = \mathbf{v}_{BA} + \mathbf{v}_{CB} + \mathbf{v}_{DC} \,.$$

Writing out the two scalar equations, we find that

$$
\begin{aligned}
0 &= -L_1\dot{\theta}_1 \sin(\theta_1) + L_2\dot{\theta}_2 \sin(\theta_2) + L_3\dot{\theta}_3 \sin(\theta_3) \,, \\
0 &= L_1\dot{\theta}_1 \cos(\theta_1) - L_2\dot{\theta}_2 \cos(\theta_2) - L_3\dot{\theta}_3 \cos(\theta_3) \,.
\end{aligned}
$$

To solve these equations for the unknown velocities $\dot{\theta}_2$ and $\dot{\theta}_3$, it is convenient to write them in matrix form:

$$
\begin{bmatrix} L_1\dot{\theta}_1 \sin(\theta_1) \\ L_1\dot{\theta}_1 \cos(\theta_1) \end{bmatrix} = \begin{bmatrix} \sin(\theta_2) & \sin(\theta_3) \\ \cos(\theta_2) & \cos(\theta_3) \end{bmatrix} \begin{bmatrix} L_2\dot{\theta}_2 \\ L_3\dot{\theta}_3 \end{bmatrix} \,.
$$

Inverting the matrix and using trigonometric identities,[9] the desired results are obtained:

$$\dot{\theta}_2 = \left(\frac{L_1 \sin(\theta_1 - \theta_3)}{L_2 \sin(\theta_2 - \theta_3)}\right)\dot{\theta}_1, \quad \dot{\theta}_3 = \left(\frac{L_1 \sin(\theta_2 - \theta_1)}{L_3 \sin(\theta_2 - \theta_3)}\right)\dot{\theta}_1.$$

This solution is valid provided that $\sin(\theta_2 - \theta_3)$ is not equal to zero. This occurs when the bars BC and DC are parallel, and in this instance it is not possible to determine the angular velocities of the bars DC and BC.

To establish equations to determine the angular accelerations of the bars DC and BC, we could differentiate the previous velocity vector equation to obtain

$$\mathbf{a}_{DA} = \mathbf{a}_{BA} + \mathbf{a}_{CB} + \mathbf{a}_{DC}.$$

The resulting 2 scalar equations, when supplemented by the 2 scalar position equations and 2 scalar velocity equations, could be used to obtain expressions for $\ddot{\theta}_2$ and $\ddot{\theta}_3$. Such a calculation would be substantial. An easier method of obtaining the desired accelerations is to differentiate the previous expressions for $\dot{\theta}_2$ and $\dot{\theta}_3$:

$$\ddot{\theta}_2 = \frac{d}{dt}\left(\frac{L_1 \sin(\theta_1 - \theta_3)}{L_2 \sin(\theta_2 - \theta_3)}\right)\dot{\theta}_1 + \left(\frac{L_1 \sin(\theta_1 - \theta_3)}{L_2 \sin(\theta_2 - \theta_3)}\right)\ddot{\theta}_1,$$

$$\ddot{\theta}_3 = \frac{d}{dt}\left(\frac{L_1 \sin(\theta_2 - \theta_1)}{L_3 \sin(\theta_2 - \theta_3)}\right)\dot{\theta}_1 + \left(\frac{L_1 \sin(\theta_2 - \theta_1)}{L_3 \sin(\theta_2 - \theta_3)}\right)\ddot{\theta}_1.$$

In summary, given the motion of the link AB, it is possible to solve for the angular displacements, angular speeds, and angular accelerations of the bars DC and BC.

The analysis of a slider crank mechanism is similar. There, the 6 unknowns are the displacement of the slider and the angular displacement of the link connecting the slider to the crank, along with the first and second time derivatives of these quantities.

8.4   Center of Mass and Linear Momentum

In all of the previous developments, we defined the motion of one material point relative to another material point of the same body. It is convenient for later purposes to now define a particular point: the center of mass C.

We first dispense with some preliminaries. Let $\mathcal{R}_0$ and $\mathcal{R}$ denote the regions of Euclidean three-space occupied by the body in its reference and

---

[9]The trigonometric identities that we use are $\sin(\alpha \pm \beta) = \sin(\alpha)\cos(\beta) \pm \sin(\beta)\cos(\alpha)$ and $\cos(\alpha \pm \beta) = \cos(\alpha)\cos(\beta) \mp \sin(\beta)\sin(\alpha)$. You may also need to recall that the inverse of the matrix $\begin{bmatrix} a & b \\ c & d \end{bmatrix}$ is $\frac{1}{ad-cb}\begin{bmatrix} d & -b \\ -c & a \end{bmatrix}$.

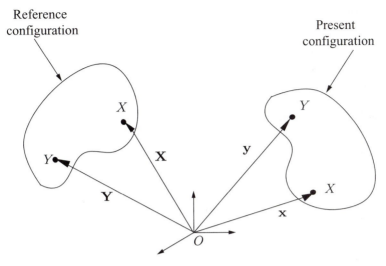

FIGURE 8.6. Position vectors of two material points X and Y of a body

present configurations, respectively. Further, let $\mathbf{X}$ and $\mathbf{x}$ be the position vectors of a material point X of the body in its reference and present configurations, respectively (see Figure 8.6).

8.4.1 The Center of Mass

The position vectors of the center of mass of the body in its reference and present configurations are defined by

$$\bar{\mathbf{X}} = \frac{\int_{\mathcal{R}_0} \mathbf{X}\rho_0 dV}{\int_{\mathcal{R}_0} \rho_0 dV} \,, \quad \bar{\mathbf{x}} = \frac{\int_{\mathcal{R}} \mathbf{x}\rho dv}{\int_{\mathcal{R}} \rho dv} \,,$$

where $\rho_0 = \rho_0(\mathbf{X})$ and $\rho = \rho(\mathbf{x}, t)$ are the mass densities per unit volume of the body in the reference and present configurations.[10]
We assume that the mass of the body is conserved:

$$dm = \rho_0 dV = \rho dv \,,$$

$$m = \int_{\mathcal{R}_0} \rho_0 dV = \int_{\mathcal{R}} \rho dv \,.$$

This is the principle of mass conservation. Hence,

$$m\bar{\mathbf{X}} = \int_{\mathcal{R}_0} \mathbf{X}\rho_0 dV \,, \quad m\bar{\mathbf{x}} = \int_{\mathcal{R}} \mathbf{x}\rho dv \,.$$

---

[10]If a body is homogeneous, then ρ_0 is a constant that is independent of $\mathbf{X}$. Our use of the symbol ρ here should not be confused with our use of the same symbol for the radius of curvature of a space curve in Chapter 3.

In addition, one has the useful identities

$$0 = \int_{\mathcal{R}_0} (\mathbf{X} - \bar{\mathbf{X}})\rho_0 dV, \quad 0 = \int_{\mathcal{R}} (\mathbf{x} - \bar{\mathbf{x}})\rho dv.$$

You should compare these expressions to those we obtained in Chapter 7 for a system of particles.

A special feature of rigid bodies is that the center of mass behaves as if it were a material point: which we denote by C. For many bodies, such as a rigid homogeneous sphere, the center of mass corresponds to the geometric center of the sphere, while for others, such as a rigid circular ring, it does not correspond to a material point. It can be proven that for any material point Y of a rigid body, one has[11]

$$\begin{bmatrix} (\bar{\mathbf{x}} - \mathbf{y}) \cdot \mathbf{E}_x \\ (\bar{\mathbf{x}} - \mathbf{y}) \cdot \mathbf{E}_y \\ (\bar{\mathbf{x}} - \mathbf{y}) \cdot \mathbf{E}_z \end{bmatrix} = \begin{bmatrix} Q_{11}(t) & Q_{12}(t) & Q_{13}(t) \\ Q_{21}(t) & Q_{22}(t) & Q_{23}(t) \\ Q_{31}(t) & Q_{32}(t) & Q_{33}(t) \end{bmatrix} \begin{bmatrix} (\bar{\mathbf{X}} - \mathbf{Y}) \cdot \mathbf{E}_x \\ (\bar{\mathbf{X}} - \mathbf{Y}) \cdot \mathbf{E}_y \\ (\bar{\mathbf{X}} - \mathbf{Y}) \cdot \mathbf{E}_z \end{bmatrix}.$$

That is $\bar{\mathbf{x}} = \chi_R(\bar{\mathbf{X}}, t)$. Differentiating these results as in Section 3, we obtain the following relations:

$$\bar{\mathbf{v}} - \dot{\mathbf{y}} = \boldsymbol{\omega} \times (\bar{\mathbf{x}} - \mathbf{y}),$$
$$\bar{\mathbf{a}} - \ddot{\mathbf{y}} = \boldsymbol{\alpha} \times (\bar{\mathbf{x}} - \mathbf{y}) + \boldsymbol{\omega} \times (\boldsymbol{\omega} \times (\bar{\mathbf{x}} - \mathbf{y})).$$

Here, $\bar{\mathbf{v}}$ and $\bar{\mathbf{a}}$ are the velocity and acceleration vectors of the center of mass C.

8.4.2   The Linear Momentum

We next turn to the linear momentum $\mathbf{G}$ of a rigid body. By definition, this momentum is

$$\mathbf{G} = \int_{\mathcal{R}} \mathbf{v}\rho dv.$$

That is, the linear momentum of a rigid body is the sum of the linear momenta of its constituents. We can establish an alternative expression for $\mathbf{G}$ using the center of mass:[12]

$$\begin{aligned} \mathbf{G} &= \int_{\mathcal{R}} \mathbf{v}\rho dv = \int_{\mathcal{R}} \frac{d\mathbf{x}}{dt}\rho dv \\ &= \frac{d}{dt}\left(\int_{\mathcal{R}} \mathbf{x}\rho dv\right) \\ &= \frac{d}{dt}(m\bar{\mathbf{x}}). \end{aligned}$$

---

[11]The proof is beyond the scope of an undergraduate engineering dynamics course. For completeness, however, one proof is discussed in Section 9.

[12]Some may notice that we take the time derivative to the outside of an integral whose region of integration $\mathcal{R}$ depends on time. This is generally not possible. However, for the integral of interest it is shown in Section 9 that such a manipulation is justified.

Hence,

$$\mathbf{G} = m\bar{\mathbf{v}}.$$

You may recall that a related result holds for a system of particles.

8.5   Kinematics of Rolling and Sliding

Many mechanical systems consist of a body in motion that is in contact with one point of another body. The resulting contact conditions are known as kinematical constraints. The discussion and description of these constraints is complicated by the fact that the particular material point of the body that is in contact changes with time.

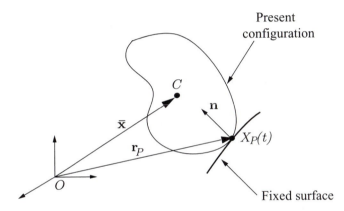

FIGURE 8.7. The geometry of contact

There are two types of contact that are of interest here: rolling and sliding. The study of rolling and sliding contact is a classical area of dynamics. In particular, some rolling rigid bodies such as the wobblestone (also known as a celt or rattleback) exhibit interesting, counterintuitive dynamics.[13]

To proceed, we consider a rigid body $\mathcal{B}$ that is in contact with a fixed surface $\mathcal{S}$ (see Figure 8.7). As the body moves on this fixed surface, the material point of the body that is in contact with the surface may change. We denote the material point of the body that is in contact at time t by $P = X_P(t)$. We denote the position vector of P by $\mathbf{r}_P$, and its velocity vector by $\mathbf{v}_P$. Finally, the unit normal to $\mathcal{S}$ at P is denoted by $\mathbf{n}$.

---

[13]The standard modern reference to this area is due to two Soviet mechanicians: Neimark and Fufaev [42]. One of the prime contributors to this area was Routh [51, 52]. Indeed, the problem of determining the motion of a sphere rolling on a surface of revolution is known as Routh's problem. We also mention the interesting classical work on billiards (pool) by Coriolis [15] from 1835.

For any material point X of $\mathcal{B}$, recall that the velocity and acceleration vectors are

$$\mathbf{v} = \bar{\mathbf{v}} + \boldsymbol{\omega} \times (\mathbf{x} - \bar{\mathbf{x}}),$$
$$\mathbf{a} = \bar{\mathbf{a}} + \boldsymbol{\alpha} \times (\mathbf{x} - \bar{\mathbf{x}}) + \boldsymbol{\omega} \times (\boldsymbol{\omega} \times (\mathbf{x} - \bar{\mathbf{x}})).$$

Consequently, for P, one has the relations

$$\mathbf{v}_P = \bar{\mathbf{v}} + \boldsymbol{\omega} \times (\mathbf{r}_P - \bar{\mathbf{x}}),$$
$$\mathbf{a}_P = \bar{\mathbf{a}} + \boldsymbol{\alpha} \times (\mathbf{r}_P - \bar{\mathbf{x}}) + \boldsymbol{\omega} \times (\boldsymbol{\omega} \times (\mathbf{r}_P - \bar{\mathbf{x}})).$$

For a rigid body that is sliding on the fixed surface $\mathcal{S}$, the component of $\mathbf{v}_P$ in the direction of $\mathbf{n}$ is zero:

$$\mathbf{v}_P \cdot \mathbf{n} = 0.$$

This implies the *sliding condition*:

$$\bar{\mathbf{v}} \cdot \mathbf{n} = -(\boldsymbol{\omega} \times (\mathbf{r}_P - \bar{\mathbf{x}})) \cdot \mathbf{n}.$$

For a rigid body that is rolling on the fixed surface $\mathcal{S}$, the velocity of the instantaneous point of contact P is zero:

$$\mathbf{v}_P = \mathbf{0}.$$

This implies the *rolling condition*:

$$\bar{\mathbf{v}} = -\boldsymbol{\omega} \times (\mathbf{r}_P - \bar{\mathbf{x}}).$$

We also note for a rolling rigid body that

$$\mathbf{a}_P = \bar{\mathbf{a}} + \boldsymbol{\alpha} \times (\mathbf{r}_P - \bar{\mathbf{x}}) + \boldsymbol{\omega} \times (\boldsymbol{\omega} \times (\mathbf{r}_P - \bar{\mathbf{x}})),$$

and this acceleration vector is not necessarily $\mathbf{0}$.

8.6   Kinematics of a Rolling Circular Disk

The main examples of rolling rigid bodies are upright rolling disks and cylinders. These examples are used to model tires in vehicle dynamics as well as numerous examples of bearing surfaces and mechanism driving devices. Here, we focus on a circular disk: the developments for a cylinder are easily inferred.[14]

---

[14]For further references to, and discussions of, rolling disks and sliding disks see Cushman, Hermans, and Kemppainen [18], Hermans [31], and O'Reilly [45]. The last of these contains a discussion of the important works on these systems by Appell and Korteweg in 1900 and Vierkandt in 1892.

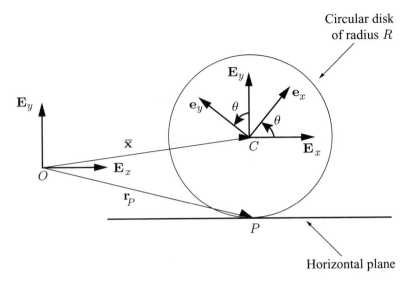

FIGURE 8.8. A circular disk rolling on a horizontal plane

As shown in Figure 8.8, we consider an upright homogeneous disk of radius R that is rolling on a plane. To start, we define a corotational basis for the disk:

$$\begin{bmatrix} \mathbf{e}_x \\ \mathbf{e}_y \\ \mathbf{e}_z \end{bmatrix} = \begin{bmatrix} \cos(\theta) & \sin(\theta) & 0 \\ -\sin(\theta) & \cos(\theta) & 0 \\ 0 & 0 & 1 \end{bmatrix} \begin{bmatrix} \mathbf{E}_x \\ \mathbf{E}_y \\ \mathbf{E}_z \end{bmatrix}.$$

Since the motion is a fixed-axis rotation,

$$\boldsymbol{\omega} = \dot{\theta}\mathbf{E}_z, \quad \boldsymbol{\alpha} = \ddot{\theta}\mathbf{E}_z.$$

Further, since the center of mass C of the disk is at its geometric center,

$$\bar{\mathbf{x}} = x\mathbf{E}_x + y\mathbf{E}_y + z\mathbf{E}_z.$$

The position vector of the instantaneous point of contact P of the disk with the plane is

$$\mathbf{r}_P = \bar{\mathbf{x}} - R\mathbf{E}_y.$$

The unit normal $\mathbf{n}$ mentioned earlier is $\mathbf{E}_y$ for this problem.

Because the disk is rolling, $\mathbf{v}_P = \mathbf{0}$, we find that

$$\begin{aligned} \bar{\mathbf{v}} &= -\boldsymbol{\omega} \times (\mathbf{r}_P - \bar{\mathbf{x}}) \\ &= -\dot{\theta}\mathbf{E}_z \times (-R\mathbf{E}_y) \\ &= -R\dot{\theta}\mathbf{E}_x. \end{aligned}$$

This vector equation is equivalent to 3 scalar equations:

$$\dot{x} = -R\dot{\theta}, \quad \dot{y} = 0, \quad \dot{z} = 0.$$

The last two of these equations imply that the velocity of C is only in the $\mathbf{E}_x$ direction, as expected. It follows from these equations that the motion of the center of mass is completely determined by the rotational motion of the disk:

$$\bar{\mathbf{v}} = \dot{x}\mathbf{E}_x = -R\dot{\theta}\mathbf{E}_x\,,$$
$$\bar{\mathbf{a}} = \ddot{x}\mathbf{E}_x = -R\ddot{\theta}\mathbf{E}_x\,.$$

Let us now examine the acceleration of the instantaneous point of contact $P = X_P(t)$. We know that the velocity vector of this point is $\mathbf{0}$. However,

$$
\begin{aligned}
\mathbf{a}_P &= \bar{\mathbf{a}} + \boldsymbol{\alpha} \times (\mathbf{r}_P - \bar{\mathbf{x}}) + \boldsymbol{\omega} \times (\boldsymbol{\omega} \times (\mathbf{r}_P - \bar{\mathbf{x}})) \\
&= \ddot{x}\mathbf{E}_x + \ddot{\theta}\mathbf{E}_z \times (-R\mathbf{E}_y) + \dot{\theta}\mathbf{E}_z \times (\dot{\theta}\mathbf{E}_z \times -R\mathbf{E}_y) \\
&= R\dot{\theta}^2\mathbf{E}_y\,.
\end{aligned}
$$

This acceleration is not zero, because the material point X_P that is in contact with the surface changes with time.

To determine the velocity and acceleration vectors of any material point X of the rolling rigid disk, we note that

$$\mathbf{x} - \bar{\mathbf{x}} = x_1\mathbf{e}_x + y_1\mathbf{e}_y\,,$$

where x_1 and y_1 are constants. Next, we use the previous results

$$
\begin{aligned}
\mathbf{v} &= \bar{\mathbf{v}} + \boldsymbol{\omega} \times (\mathbf{x} - \bar{\mathbf{x}})\,, \\
\mathbf{a} &= \bar{\mathbf{a}} + \boldsymbol{\alpha} \times (\mathbf{x} - \bar{\mathbf{x}}) + \boldsymbol{\omega} \times (\boldsymbol{\omega} \times (\mathbf{x} - \bar{\mathbf{x}}))\,.
\end{aligned}
$$

Since,

$$\bar{\mathbf{v}} = -R\dot{\theta}\mathbf{E}_x\,, \quad \bar{\mathbf{a}} = -R\ddot{\theta}\mathbf{E}_x\,, \quad \boldsymbol{\omega} = \dot{\theta}\mathbf{E}_z\,, \quad \boldsymbol{\alpha} = \ddot{\theta}\mathbf{E}_z\,,$$

we find that

$$
\begin{aligned}
\mathbf{v} &= -R\dot{\theta}\mathbf{E}_x + \dot{\theta}\mathbf{E}_z \times (x_1\mathbf{e}_x + y_1\mathbf{e}_y) \\
&= -R\dot{\theta}\mathbf{E}_x + \dot{\theta}(x_1\mathbf{e}_y - y_1\mathbf{e}_x)\,, \\
\mathbf{a} &= -R\ddot{\theta}\mathbf{E}_x + \ddot{\theta}\mathbf{E}_z \times (x_1\mathbf{e}_x + y_1\mathbf{e}_y) + \dot{\theta}\mathbf{E}_z \times (\dot{\theta}\mathbf{E}_z \times (x_1\mathbf{e}_x + y_1\mathbf{e}_y)) \\
&= -R\ddot{\theta}\mathbf{E}_x + \ddot{\theta}(x_1\mathbf{e}_y - y_1\mathbf{e}_x) - \dot{\theta}^2(x_1\mathbf{e}_x + y_1\mathbf{e}_y)\,.
\end{aligned}
$$

It is a good exercise to choose various values of x_1 and y_1 and examine the corresponding velocity and acceleration vectors.

8.6.1 A Common Error

It is a commmon error to start with the equation $\mathbf{r}_P = \bar{\mathbf{x}} - R\mathbf{E}_y$, and then differentiate this equation to try to get $\mathbf{v}_P$ and $\mathbf{a}_P$. One gets the incorrect answers $\mathbf{v}_P = \bar{\mathbf{v}}$ and $\mathbf{a}_P = \bar{\mathbf{a}}$ when one does this.

The obvious question is why this occurs. The reason is that the position vector of $X_P(t)$ relative to the center of mass C is $-R\mathbf{E}_y$ only at the instant t. At other times, its relative position vector does not have this value. When one differentiates $\mathbf{r}_P = \bar{\mathbf{x}} - R\mathbf{E}_y$, the derivative of $R\mathbf{E}_y$ is equal to $\mathbf{0}$. Hence, one is falsely assuming that the same material point is the instantaneous point of contact for an interval of time.

8.6.2   The Sliding Disk

It is interesting to pause briefly to consider the sliding disk. For such a disk, the sliding condition yields

$$\bar{\mathbf{v}} \cdot \mathbf{E}_y = -\left(\dot{\theta}\mathbf{E}_z \times (-R\mathbf{E}_y)\right) \cdot \mathbf{E}_y\,.$$

This implies that

$$\bar{\mathbf{v}} \cdot \mathbf{E}_y = \dot{y} = 0\,.$$

Hence, the rotational and translational motions of the disk are not coupled.

8.7   Angular Momenta

Preparatory to a discussion of the balance laws for a rigid body, we now address the angular momenta of a rigid body. The momentum relative to two points, the center of mass C and a fixed point O, will be of considerable importance in the next two chapters. For convenience, we shall assume that the fixed point O is also the origin (see Figure 8.9).

By definition, the angular momenta of a rigid body relative to its center of mass C, $\mathbf{H}$, and a fixed point O, $\mathbf{H}_O$, are

$$\mathbf{H} = \int_{\mathcal{R}} (\mathbf{x} - \bar{\mathbf{x}}) \times \mathbf{v}\rho dv\,, \quad \mathbf{H}_O = \int_{\mathcal{R}} \mathbf{x} \times \mathbf{v}\rho dv\,.$$

These momenta are related by a simple and important formula. To find this formula, we perform some manipulations on $\mathbf{H}_O$:

$$
\begin{aligned}
\mathbf{H}_O &= \int_{\mathcal{R}} \mathbf{x} \times \mathbf{v}\rho dv \\
&= \int_{\mathcal{R}} (\mathbf{x} - \bar{\mathbf{x}} + \bar{\mathbf{x}}) \times \mathbf{v}\rho dv \\
&= \int_{\mathcal{R}} (\mathbf{x} - \bar{\mathbf{x}}) \times \mathbf{v}\rho dv + \int_{\mathcal{R}} \bar{\mathbf{x}} \times \mathbf{v}\rho dv \\
&= \mathbf{H} + \bar{\mathbf{x}} \times \int_{\mathcal{R}} \mathbf{v}\rho dv\,.
\end{aligned}
$$

That is,

$$\mathbf{H}_O = \mathbf{H} + \bar{\mathbf{x}} \times \mathbf{G}\,.$$

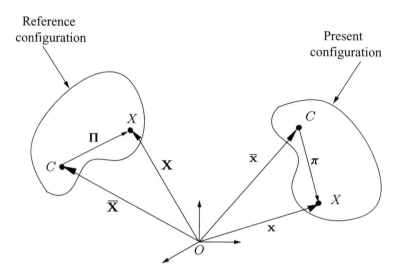

FIGURE 8.9. Relative position vectors $\boldsymbol{\Pi}$ and $\boldsymbol{\pi}$ of a material point X of a body

In words, the angular momentum of a rigid body relative to a fixed point O is the sum of the angular momentum of the rigid body about its center of mass and the angular momentum of its center of mass relative to O.

As an extension of the previous result, it is a good exercise to show that the angular momentum of a rigid body relative to an arbitrary point A satisfies the identity

$$\mathbf{H}_A = \int_{\mathcal{R}} (\mathbf{x} - \mathbf{x}_A) \times \mathbf{v}\rho dv = \mathbf{H} + (\bar{\mathbf{x}} - \mathbf{x}_A) \times \mathbf{G}.$$

Here, $\mathbf{x}_A$ is the position vector of the point A. You might recall that we had a similar identity for a system of particles.

In the forthcoming balance laws, we will need to measure $\mathbf{H}$ and $\mathbf{H}_O$ at various instants of time. Using the formulae above, this will be a tedious task. It is simplified tremendously by the introduction of inertia tensors.

8.8   Inertia Tensors

In this section we first establish the inertia tensor for a rigid body relative to its center of mass C. Some comments on the parallel-axis theorem are presented at the end of this section.

To start, we recall that for any material point X of a rigid body, one has the relation

$$\mathbf{v} = \bar{\mathbf{v}} + \boldsymbol{\omega} \times (\mathbf{x} - \bar{\mathbf{x}}).$$

As shown in Figure 8.9, we also define the relative position vectors

$$\boldsymbol{\pi} = \mathbf{x} - \bar{\mathbf{x}}, \quad \boldsymbol{\Pi} = \mathbf{X} - \bar{\mathbf{X}}.$$

Because the motion of the body is rigid, these vectors have the following representations and satisfy the following identities:

$$\boldsymbol{\pi} \;=\; \mathbf{x} - \bar{\mathbf{x}} = \Pi_x \mathbf{e}_x + \Pi_y \mathbf{e}_y + \Pi_z \mathbf{e}_z\,,$$
$$\boldsymbol{\Pi} \;=\; \mathbf{X} - \bar{\mathbf{X}} = \Pi_x \mathbf{E}_x + \Pi_y \mathbf{E}_y + \Pi_z \mathbf{E}_z\,.$$

Notice that the components of $\boldsymbol{\Pi}$ relative to the fixed Cartesian basis are identical to those of $\boldsymbol{\pi}$ relative to the corotational basis. This implies that the latter components can be measured using the fixed reference configuration.

8.8.1 Where the Inertia Tensor Comes From

Consider the angular momentum $\mathbf{H}$:

$$
\begin{aligned}
\mathbf{H} \;&=\; \int_{\mathcal{R}} (\mathbf{x} - \bar{\mathbf{x}}) \times \mathbf{v} \rho \, dv \\
&=\; \int_{\mathcal{R}} \boldsymbol{\pi} \times \mathbf{v} \rho \, dv \\
&=\; \int_{\mathcal{R}} \boldsymbol{\pi} \times (\bar{\mathbf{v}} + \boldsymbol{\omega} \times \boldsymbol{\pi}) \rho \, dv \\
&=\; \int_{\mathcal{R}} \boldsymbol{\pi} \times \bar{\mathbf{v}} \rho \, dv + \int_{\mathcal{R}} \boldsymbol{\pi} \times (\boldsymbol{\omega} \times \boldsymbol{\pi}) \rho \, dv \,.
\end{aligned}
$$

However, since C is the center of mass and the velocity vector of C is independent of the region of integration,

$$\int_{\mathcal{R}} \boldsymbol{\pi} \times \bar{\mathbf{v}} \rho \, dv = \int_{\mathcal{R}} \boldsymbol{\pi} \rho \, dv \times \bar{\mathbf{v}} = \mathbf{0} \times \bar{\mathbf{v}} = \mathbf{0}\,.$$

Hence,

$$\mathbf{H} = \int_{\mathcal{R}} \boldsymbol{\pi} \times (\boldsymbol{\omega} \times \boldsymbol{\pi}) \rho \, dv = \int_{\mathcal{R}} \big((\boldsymbol{\pi} \cdot \boldsymbol{\pi}) \boldsymbol{\omega} - (\boldsymbol{\pi} \cdot \boldsymbol{\omega}) \boldsymbol{\pi} \big) \, \rho \, dv\,.$$

In writing this equation, we used the identity $\mathbf{a} \times (\mathbf{b} \times \mathbf{c}) = (\mathbf{a} \cdot \mathbf{c}) \mathbf{b} - (\mathbf{a} \cdot \mathbf{b}) \mathbf{c}$. Substituting the representations

$$\boldsymbol{\pi} \;=\; \mathbf{x} - \bar{\mathbf{x}} = \Pi_x \mathbf{e}_x + \Pi_y \mathbf{e}_y + \Pi_z \mathbf{e}_z\,,$$
$$\boldsymbol{\omega} \;=\; \omega_x \mathbf{e}_x + \omega_y \mathbf{e}_y + \omega_z \mathbf{e}_z\,,$$

into the last equation, and expanding, we find that

$$
\begin{aligned}
\mathbf{H} \;=\; & \left(I_{xx}\omega_x + I_{xy}\omega_y + I_{xz}\omega_z \right) \mathbf{e}_x \\
& + \left(I_{xy}\omega_x + I_{yy}\omega_y + I_{yz}\omega_z \right) \mathbf{e}_y \\
& + \left(I_{xz}\omega_x + I_{yz}\omega_y + I_{zz}\omega_z \right) \mathbf{e}_z \,.
\end{aligned}
$$

In this expression, the inertias $I_{xx}, \ldots, I_{zz}$ are the 6 independent components of the inertia tensor of the body relative to its center of mass:

$$I_{xx} = \int_{\mathcal{R}} \left(\Pi_y^2 + \Pi_z^2 \right) \rho dv = \int_{\mathcal{R}_0} \left(\Pi_y^2 + \Pi_z^2 \right) \rho_0 dV \, ,$$

$$I_{xy} = -\int_{\mathcal{R}} \Pi_x \Pi_y \rho dv = -\int_{\mathcal{R}_0} \Pi_x \Pi_y \rho_0 dV \, ,$$

$$I_{xz} = -\int_{\mathcal{R}} \Pi_x \Pi_z \rho dv = -\int_{\mathcal{R}_0} \Pi_x \Pi_z \rho_0 dV \, ,$$

$$I_{yy} = \int_{\mathcal{R}} \left(\Pi_x^2 + \Pi_z^2 \right) \rho dv = \int_{\mathcal{R}_0} \left(\Pi_x^2 + \Pi_z^2 \right) \rho_0 dV \, ,$$

$$I_{yz} = -\int_{\mathcal{R}} \Pi_y \Pi_z \rho dv = -\int_{\mathcal{R}_0} \Pi_y \Pi_z \rho_0 dV \, ,$$

$$I_{zz} = \int_{\mathcal{R}} \left(\Pi_x^2 + \Pi_y^2 \right) \rho dv = \int_{\mathcal{R}_0} \left(\Pi_x^2 + \Pi_y^2 \right) \rho_0 dV \, .$$

You should notice that all of the inertias can be evaluated in the fixed reference configuration of the rigid body.[15]

The integrals in the expressions for $I_{xx}, \ldots, I_{zz}$ are standard volume integrals. For many bodies they are tabulated in texts, for example, Table D/4 of Meriam and Kraige [39]. In most texts, I_{xx}, I_{yy}, and I_{zz} are known as the moments of inertia, while $-I_{xy}$, $-I_{xz}$, and $-I_{yz}$, are known as the products of inertia.

8.8.2 Angular Momentum and the Inertia Tensor

Recall that we have just shown that

$$\mathbf{H} = (I_{xx}\omega_x + I_{xy}\omega_y + I_{xz}\omega_z)\,\mathbf{e}_x$$
$$+ (I_{xy}\omega_x + I_{yy}\omega_y + I_{yz}\omega_z)\,\mathbf{e}_y$$
$$+ (I_{xz}\omega_x + I_{yz}\omega_y + I_{zz}\omega_z)\,\mathbf{e}_z \, .$$

We can write this result in a more transparent form:

$$\begin{bmatrix} \mathbf{H} \cdot \mathbf{e}_x \\ \mathbf{H} \cdot \mathbf{e}_y \\ \mathbf{H} \cdot \mathbf{e}_z \end{bmatrix} = \begin{bmatrix} I_{xx} & I_{xy} & I_{xz} \\ I_{xy} & I_{yy} & I_{yz} \\ I_{xz} & I_{yz} & I_{zz} \end{bmatrix} \begin{bmatrix} \boldsymbol{\omega} \cdot \mathbf{e}_x \\ \boldsymbol{\omega} \cdot \mathbf{e}_y \\ \boldsymbol{\omega} \cdot \mathbf{e}_z \end{bmatrix} .$$

The matrix in this equation is known as the inertia matrix. Its components are the components of the inertia tensor.

---

[15]These results are discussed in Casey [11, 14] and in Section 13, Chapter 4 of Gurtin [30]. For further details on the transformation of the integrals, see Section 9 below.

We notice that the inertia matrix is symmetric. It may also be shown that it is positive definite. As a result, its eigenvalues (or principal values) are real positive numbers.[16] You should also notice that the components of this matrix depend on the corotational basis chosen, and, as a result, the fixed Cartesian basis also. If the vectors $\mathbf{E}_x$, $\mathbf{E}_y$, and $\mathbf{E}_z$ are chosen to coincide with the eigenvectors of this matrix, then the sets of vectors $\{\mathbf{E}_x, \mathbf{E}_y, \mathbf{E}_z\}$ and $\{\mathbf{e}_x, \mathbf{e}_y, \mathbf{e}_z\}$ are said to be the principal axes of the body in its reference and present configurations, respectively. In this case, the above expression for $\mathbf{H}$ simplifies considerably:

$$
\begin{bmatrix} \mathbf{H} \cdot \mathbf{e}_x \\ \mathbf{H} \cdot \mathbf{e}_y \\ \mathbf{H} \cdot \mathbf{e}_z \end{bmatrix} = \begin{bmatrix} I_{xx} & 0 & 0 \\ 0 & I_{yy} & 0 \\ 0 & 0 & I_{zz} \end{bmatrix} \begin{bmatrix} \boldsymbol{\omega} \cdot \mathbf{e}_x \\ \boldsymbol{\omega} \cdot \mathbf{e}_y \\ \boldsymbol{\omega} \cdot \mathbf{e}_z \end{bmatrix} .
$$

If possible, one chooses $\{\mathbf{E}_x, \mathbf{E}_y, \mathbf{E}_z\}$ and $\{\mathbf{e}_x, \mathbf{e}_y, \mathbf{e}_z\}$ to be the sets of principal axes.

8.8.3   A Circular Cylinder

As an example, we consider a rigid homogeneous circular cylinder of mass m, radius R, and length L shown in Figure 8.10. For this body, we choose $\{\mathbf{E}_x, \mathbf{E}_y, \mathbf{E}_z\}$ to be the principal axes of the body: $I_{xy} = I_{xz} = I_{yz} = 0$. The center of mass C of the cylinder is at its geometric center.

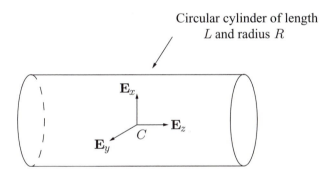

Circular cylinder of length
L and radius R

FIGURE 8.10. A circular cylinder of mass m, radius R, and length L

Evaluating the volume integrals discussed previously, one obtains

$$
\begin{bmatrix} I_{xx} & I_{xy} & I_{xz} \\ I_{xy} & I_{yy} & I_{yz} \\ I_{xz} & I_{yz} & I_{zz} \end{bmatrix} = \begin{bmatrix} \frac{1}{4}mR^2 + \frac{1}{12}mL^2 & 0 & 0 \\ 0 & \frac{1}{4}mR^2 + \frac{1}{12}mL^2 & 0 \\ 0 & 0 & \frac{1}{2}mR^2 \end{bmatrix} .
$$

---

[16]These results are well-known in linear algebra, see, for example, Bellman [8] or Strang [61]. One also uses these results for the (Cauchy) stress tensor when constructing Mohr's circle in solid mechanics courses.

Notice that by setting $R = 0$ you can use these results to determine the inertias of a slender rod. Similarly, by setting $L = 0$, the inertias for a thin circular disk can be obtained

8.8.4   The Parallel-Axis Theorem

In many problems, it is convenient to consider the inertia matrix relative to a point A which is not the center of mass. The relevant inertias $I_{Axx}, \ldots, I_{Azz}$ can be determined from the components of the inertia matrix for the body relative to its center of mass using the parallel-axis theorem (see, for instance, Appendix B of Meriam and Kraige [39]). When A is a point of the rigid body, the inertias $I_{Axx}, \ldots, I_{Azz}$ can then be used to determine a convenient expression for $\mathbf{H}_A$. Here, we find it more convenient to use the relationship

$$\mathbf{H}_A = \int_{\mathcal{R}} (\mathbf{x} - \mathbf{x}_A) \times \mathbf{v} \rho dv = \mathbf{H} + (\bar{\mathbf{x}} - \mathbf{x}_A) \times \mathbf{G}.$$

Here, $\mathbf{x}_A$ is the position vector of the point A. This equation holds even if A is not a material point of the rigid body. Moreover, it subsumes the parallel-axis theorem. A specific example is discussed in Section 4 of Chapter 9.

8.9   Some Comments on Integrals and Derivatives

In several of the developments in Sections 4, 7, and 8 some advanced identities were used. These are beyond the scope of an undergraduate engineering dynamics course. Here, we present some further details on these identities for the interested reader.

A difficulty with evaluating the expressions for $\bar{\mathbf{x}}$ and $\mathbf{H}$ is that the region of integration $\mathcal{R}$ depends on time. We shall have similar issues in the next chapter when evaluating the derivatives of certain integrals. We now record some results pertaining to these integrals when the motion of the body is rigid: $\mathbf{x} = \boldsymbol{\chi}_R(\mathbf{X}, t)$. Proofs of these results can be found in the literature on continuum mechanics.[17]

First, we have the local conservation of mass result

$$\rho_0(\mathbf{X}) = \rho(\mathbf{x} = \boldsymbol{\chi}_R(\mathbf{X}, t), t).$$

In words, this implies that the mass density at a material point X of the rigid body is the same in its reference and present configurations.

---

[17]These results follow from the local forms of mass conservation, changes of variables theorem, and Reynolds' transport theorem in continuum mechanics because a rigid body's motion is isochoric, see, for example, Casey [11], Gurtin [30], and Truesdell and Toupin [66].

The second result is a change of variables theorem for any sufficiently smooth function f:

$$\int_{\mathcal{R}} f(\mathbf{x}, t) dv = \int_{\mathcal{R}_0} f(\boldsymbol{\chi}_R(\mathbf{X}, t), t) dV .$$

We used this result and mass conservation to establish expressions for the inertias in Section 8.

The last result is a version of Reynolds' transport theorem for a sufficiently smooth function g:

$$\frac{d}{dt} \int_{\mathcal{R}} g(\mathbf{x}, t) \rho dv = \int_{\mathcal{R}} \frac{d}{dt}(g(\mathbf{x}, t)) \rho dv .$$

We used this result in Section 4 when we took the time derivative outside the integral to establish the result that $\mathbf{G} = m\bar{\mathbf{v}}$.

In Section 4 we used the result that the center of mass of a rigid body behaves as if it were a material point. This result is accepted without comment by most texts. The first proof of it, to our knowledge, is by Casey [11]. Here, we outline his proof. Our outline is far longer than his proof. The reason for this is that he uses a compact tensor notation. We first recall, from Section 1, that for any two material points X and Y of a rigid body,

$$\begin{bmatrix} (\mathbf{x} - \mathbf{y}) \cdot \mathbf{E}_x \\ (\mathbf{x} - \mathbf{y}) \cdot \mathbf{E}_y \\ (\mathbf{x} - \mathbf{y}) \cdot \mathbf{E}_z \end{bmatrix} = \begin{bmatrix} Q_{11}(t) & Q_{12}(t) & Q_{13}(t) \\ Q_{21}(t) & Q_{22}(t) & Q_{23}(t) \\ Q_{31}(t) & Q_{32}(t) & Q_{33}(t) \end{bmatrix} \begin{bmatrix} (\mathbf{X} - \mathbf{Y}) \cdot \mathbf{E}_x \\ (\mathbf{X} - \mathbf{Y}) \cdot \mathbf{E}_y \\ (\mathbf{X} - \mathbf{Y}) \cdot \mathbf{E}_z \end{bmatrix} .$$

One now substitutes these results into the right-hand side of the identity

$$\bar{\mathbf{x}} - \mathbf{y} = \frac{1}{m} \int_{\mathcal{R}} \mathbf{x} \rho dv - \mathbf{y} = \frac{1}{m} \int_{\mathcal{R}} \mathbf{x} \rho dv - \frac{\int_{\mathcal{R}} \rho dv}{m} \mathbf{y} = \frac{1}{m} \int_{\mathcal{R}} (\mathbf{x} - \mathbf{y}) \rho dv .$$

Next, one uses the change of variables result recorded above and the definition of $\bar{\mathbf{X}}$ to conclude that

$$\begin{bmatrix} (\bar{\mathbf{x}} - \mathbf{y}) \cdot \mathbf{E}_x \\ (\bar{\mathbf{x}} - \mathbf{y}) \cdot \mathbf{E}_y \\ (\bar{\mathbf{x}} - \mathbf{y}) \cdot \mathbf{E}_z \end{bmatrix} = \begin{bmatrix} Q_{11}(t) & Q_{12}(t) & Q_{13}(t) \\ Q_{21}(t) & Q_{22}(t) & Q_{23}(t) \\ Q_{31}(t) & Q_{32}(t) & Q_{33}(t) \end{bmatrix} \begin{bmatrix} (\bar{\mathbf{X}} - \mathbf{Y}) \cdot \mathbf{E}_x \\ (\bar{\mathbf{X}} - \mathbf{Y}) \cdot \mathbf{E}_y \\ (\bar{\mathbf{X}} - \mathbf{Y}) \cdot \mathbf{E}_z \end{bmatrix} .$$

This result implies that the center of mass C of the rigid body behaves as if it were a material point of the body. For deformable bodies, this is not true.[18]

---

[18]For example, take a flexible ruler. Initially, suppose that it is straight. One can approximately locate its center of mass; suppose that it is at the geometric center. Now, bend the ruler into a circle. The center of mass no longer coincides with the same material point of the ruler.

8.10   Summary

The first part of this chapter was devoted to examining the kinematical relationships between the position vectors, velocity vectors and acceleration vectors of two material points X_1 and X_2 of a rigid body. Then expressions for the linear momentum $\mathbf{G}$, and angular momenta $\mathbf{H}$, $\mathbf{H}_O$, and $\mathbf{H}_A$ were presented. Expressions for the angular momenta were simplified using the inertia tensor. For pedagogical reasons, we found it convenient to present many of the results for arbitrary rotations and then simplify them for the case of a fixed-axis of rotation.

Denoting the position vectors of the material points X_1 and X_2 by $\mathbf{x}_1$ and $\mathbf{x}_2$, respectively, it was shown that

$$
\begin{aligned}
\mathbf{v}_1 - \mathbf{v}_2 &= \boldsymbol{\omega} \times (\mathbf{x}_1 - \mathbf{x}_2)\,, \\
\mathbf{a}_1 - \mathbf{a}_2 &= \boldsymbol{\alpha} \times (\mathbf{x}_1 - \mathbf{x}_2) + \boldsymbol{\omega} \times (\boldsymbol{\omega} \times (\mathbf{x}_1 - \mathbf{x}_2))\,,
\end{aligned}
$$

where $\boldsymbol{\omega} = \omega_x \mathbf{e}_x + \omega_y \mathbf{e}_y + \omega_z \mathbf{e}_z$ is the angular velocity vector of the rigid body and $\boldsymbol{\alpha} = \dot{\boldsymbol{\omega}}$ is the angular acceleration vector of the rigid body. To facilitate working with several problems, a corotational basis $\{\mathbf{e}_x, \mathbf{e}_y, \mathbf{e}_z\}$ was also introduced. This basis rotates with the body, and we also showed that $\dot{\mathbf{e}}_x = \boldsymbol{\omega} \times \mathbf{e}_x$, $\dot{\mathbf{e}}_y = \boldsymbol{\omega} \times \mathbf{e}_y$, and $\dot{\mathbf{e}}_z = \boldsymbol{\omega} \times \mathbf{e}_z$. In Section 4.1, the center of mass C of the rigid body was introduced. For a rigid body, C behaves as a material point of the rigid body. In addition, the linear momentum of the rigid body is

$$
\mathbf{G} = m\bar{\mathbf{v}}\,,
$$

where $\bar{\mathbf{v}}$ is the velocity vector of the center of mass and m is the mass of the rigid body. The angular momentum of a rigid body relative to its center of mass has the representation

$$
\begin{aligned}
\mathbf{H} =\ & (I_{xx}\omega_x + I_{xy}\omega_y + I_{xz}\omega_z)\,\mathbf{e}_x \\
& + (I_{xy}\omega_x + I_{yy}\omega_y + I_{yz}\omega_z)\,\mathbf{e}_y \\
& + (I_{xz}\omega_x + I_{yz}\omega_y + I_{zz}\omega_z)\,\mathbf{e}_z\,.
\end{aligned}
$$

The inertias $I_{xx}, \ldots, I_{zz}$ are constants associated with the rigid body. They depend on the mass and geometry of the rigid body and the choice of the basis vectors $\{\mathbf{E}_x, \mathbf{E}_y, \mathbf{E}_z\}$. If possible these vectors are chosen to be principal axes of the body, in which case $I_{xy} = I_{yz} = I_{xz} = 0$. The following relationships were also established:

$$
\mathbf{H}_O = \mathbf{H} + \bar{\mathbf{x}} \times \mathbf{G}\,, \quad \mathbf{H}_A = \mathbf{H} + (\bar{\mathbf{x}} - \mathbf{x}_A) \times \mathbf{G}\,.
$$

Most of the results in this chapter were specialized to the case of a fixed-axis of rotation. This axis was chosen to be $\mathbf{E}_z$, and the angle of rotation of the rigid body was denoted by θ. All of the aforementioned kinematical results simplify for this case. For instance,

$$
\boldsymbol{\omega} = \dot{\theta}\mathbf{E}_z\,, \quad \boldsymbol{\alpha} = \ddot{\theta}\mathbf{E}_z\,,
$$

and

$$\mathbf{e}_x = \cos(\theta)\mathbf{E}_x + \sin(\theta)\mathbf{E}_y\,, \quad \mathbf{e}_y = \cos(\theta)\mathbf{E}_y - \sin(\theta)\mathbf{E}_x\,, \quad \mathbf{e}_z = \mathbf{E}_z\,.$$

The most substantial simplification occurs in the expression for $\mathbf{H}$:

$$\mathbf{H} = I_{xz}\dot{\theta}\mathbf{e}_x + I_{yz}\dot{\theta}\mathbf{e}_y + I_{zz}\dot{\theta}\mathbf{E}_z\,.$$

The kinematical results presented in the chapter were used to examine the kinematics of mechanisms and rolling and sliding rigid bodies. In the mechanism problems, it was shown how the determine the angular velocities and angular accelerations of two of the linkages in a four-bar mechanism as functions of the angular velocity and acceleration of a third linkage. For rolling and sliding rigid bodies, the important conditions $\mathbf{v}_P = \mathbf{0}$ and $\mathbf{v}_P \cdot \mathbf{n} = 0$ were discussed.

8.11   Exercises

The following short exercises are intended to assist you in reviewing Chapter 8. In the exercises, we restrict attention to the fixed-axis of rotation case: $\boldsymbol{\omega} = \dot{\theta}\mathbf{E}_z$.

8.1 Show that
$$\dot{\mathbf{e}}_x = \dot{\theta}\mathbf{e}_y\,, \quad \dot{\mathbf{e}}_y = -\dot{\theta}\mathbf{e}_x\,, \quad \dot{\mathbf{e}}_z = \mathbf{0}\,.$$

8.2 The position vector of a material point of a rigid body relative to a fixed point O of the rigid body is

$$\mathbf{x} = x\mathbf{e}_x + y\mathbf{e}_y + z\mathbf{e}_z\,.$$

By differentiating this result, show that the expressions for $\mathbf{v}$ and $\mathbf{a}$ are identical to those you would have obtained had you used the formulae
$$\mathbf{v} = \boldsymbol{\omega} \times \mathbf{x}\,, \quad \mathbf{a} = \boldsymbol{\alpha} \times \mathbf{x} + \boldsymbol{\omega} \times (\boldsymbol{\omega} \times \mathbf{x})\,.$$

8.3 A particle of mass m is free to move in a groove which is machined on a rigid body. The center of mass C of the rigid body is fixed. If the position vector of the particle relative to C is

$$\mathbf{r} - \bar{\mathbf{x}} = 5\mathbf{e}_x + y\mathbf{e}_y\,,$$

calculate $\dot{\mathbf{r}}$ and $\ddot{\mathbf{r}}$. Here, y is not a constant.

8.4 How do the results of Exercise 8.3 change if the center of mass C has a motion $\bar{\mathbf{x}} = 10t\mathbf{E}_x + t^2\mathbf{E}_y$?

8.5 For the mechanism discussed in Section 3, determine the angular velocities and accelerations of the links DC and BC if

$$L_1 = 10 \, \text{meters}, \quad L_2 = 5 \, \text{meters}, \quad L_3 = 10 \, \text{meters},$$

and

$$\theta_1 = \frac{\pi}{2} \, \text{radians}, \quad \dot{\theta}_1 = 1.0 \, \text{RPM}, \quad \ddot{\theta}_1 = 0.1 \, \text{radians/sec}^2.$$

8.6 For the rolling disk discussed in Section 6, determine the velocity and acceleration vectors of the material point X whose position vector relative to the center of mass is

$$\mathbf{x} - \bar{\mathbf{x}} = R\mathbf{e}_x.$$

At an instant during each revolution of the disk, why does X have a velocity vector $\mathbf{0}$?

8.7 For the sliding disk discussed in Section 6.2, determine the velocity and acceleration vectors of the material point X whose position vector relative to the center of mass is

$$\mathbf{x} - \bar{\mathbf{x}} = R\mathbf{e}_x.$$

When is it possible for X to have the same acceleration and velocity as the center of mass?

8.8 For the rolling conditions on a circular disk discussed in Section 6, many people, in haste, write $\dot{x} = R\dot{\theta}$. Show that this result does not imply that $\mathbf{v}_P \cdot \mathbf{E}_x = 0$. What does it imply?

8.9 For the rolling circular disk discussed in Section 6, derive expressions for $\mathbf{H}$ and $\mathbf{H}_P$.

8.10 Calculate the time derivative of

$$\mathbf{H} = I_{xz}\dot{\theta}\mathbf{e}_x + I_{yz}\dot{\theta}\mathbf{e}_y + I_{zz}\dot{\theta}\mathbf{E}_z.$$

The answer is displayed in Section 1.3 of Chapter 9.

9
Kinetics of a Rigid Body

TOPICS

We start by discussing Euler's laws for a rigid body. These laws are known as the balances of linear and angular momenta. An alternative form of these laws is also presented that is useful for solving problems. We then discuss the kinetic energy of a rigid body and establish the Koenig decomposition. This decomposition, combined with the balance laws, can be used to prove a work-energy theorem for a rigid body. As illustrations of the theory we consider four classes of problems: purely translational motion of a rigid body, rigid bodies that are free to rotate about one of their fixed material points, rolling and sliding bodies, and an imbalanced rotor problem. These applications are far from exhaustive, but we feel they are the chief representatives of problems for an undergraduate engineering dynamics course.

9.1   Balance Laws for a Rigid Body

Euler's laws for a rigid body can be viewed as extensions to Newton's second law for a particle. There are two laws, or postulates, the balance of linear momentum and the balance of angular momentum.[1]

---

[1] For a single particle or system of particles, the balance of angular momentum is not an independent postulate: as shown in Chapter 7, it follows from the balance of linear momentum. You should also notice that we do not attempt, as many texts do, to derive the balances of linear and angular momenta from the balance of linear momentum for

9.1.1   Resultant Forces and Moments

Before discussing the balance laws, we dispense with some preliminaries. The resultant force $\mathbf{F}$ acting on the rigid body is the sum of all the forces acting on the rigid body. Similarly, the resultant moment relative to a fixed point O, $\mathbf{M}_O$, is the resultant external moment relative to O of all of the moments acting on the rigid body. We also denote the resultant moment relative to the center of mass C by $\mathbf{M}$. These moments may be decomposed into two additive parts, the moment due to the individual external forces acting on the rigid body and applied external moments that are not due to external forces.

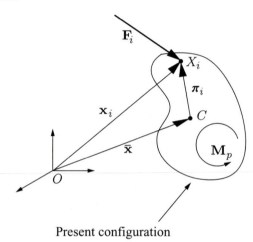

Present configuration

FIGURE 9.1. A force $\mathbf{F}_i$ and a moment $\mathbf{M}_p$ acting on a rigid body

As an example, consider a system of forces and moments acting on a rigid body. Here, a set of K forces $\mathbf{F}_i$ $(i = 1, \ldots, K)$ act on the rigid body. The force $\mathbf{F}_i$ acts at the material point X_i, which has a position vector $\mathbf{x}_i$. In addition, a moment $\mathbf{M}_p$, which is not due to the moment of an applied force, also acts on the rigid body (see Figure 9.1). For this system of applied forces and moments, the resultants are

$$\mathbf{F} = \sum_{i=1}^{K} \mathbf{F}_i,$$

$$\mathbf{M}_O = \mathbf{M}_p + \sum_{i=1}^{K} \mathbf{x}_i \times \mathbf{F}_i,$$

---

each of the material points of a rigid body. Such a derivation entails placing restrictions on the nature of the internal forces acting on the particles. Here, we follow the approach in continuum mechanics, and postulate two independent balance laws (see Essay V of Truesdell [65] for further discussions on this matter).

$$\mathbf{M} \; = \; \mathbf{M}_p + \sum_{i=1}^{K} (\mathbf{x}_i - \bar{\mathbf{x}}) \times \mathbf{F}_i \, .$$

Notice how $\mathbf{M}_p$ features in these expressions.

9.1.2   Euler's Laws

The balance laws for a rigid body are often known as Euler's laws.[2] The first of these laws is the balance of linear momentum:

$$\mathbf{F} = \dot{\mathbf{G}} = m\dot{\mathbf{v}} \, .$$

The second law is the balance of angular momentum for a rigid body is

$$\mathbf{M}_O = \dot{\mathbf{H}}_O \, .$$

These balance laws represent 6 scalar equations.

In many cases it is convenient to give an alternative description of the balance of angular momentum. To do this we start with the identity

$$\mathbf{H}_O = \mathbf{H} + \bar{\mathbf{x}} \times \mathbf{G} \, .$$

Differentiating, and using the balance of linear momentum, we find that

$$\begin{aligned} \dot{\mathbf{H}}_O \; &= \; \dot{\mathbf{H}} + \bar{\mathbf{v}} \times \mathbf{G} + \bar{\mathbf{x}} \times \dot{\mathbf{G}} \\ &= \; \dot{\mathbf{H}} + \bar{\mathbf{x}} \times \mathbf{F} \, . \end{aligned}$$

Hence, invoking the balance of angular momentum,

$$\mathbf{M}_O = \dot{\mathbf{H}}_O = \dot{\mathbf{H}} + \bar{\mathbf{x}} \times \mathbf{F} \, .$$

However, the resultant moment relative to a fixed point O, $\mathbf{M}_O$, and the resultant moment relative to the center of mass C, $\mathbf{M}$, are related by[3]

$$\mathbf{M}_O = \mathbf{M} + \bar{\mathbf{x}} \times \mathbf{F} \, .$$

It follows that

$$\mathbf{M} = \dot{\mathbf{H}} \, ,$$

which is known as the balance of angular momentum relative to the center of mass C. This form of the balance law is used in many problems where the rigid body has no fixed point O.

---

[2]See Truesdell [65]: For a rigid body, they may be seen on pages 224–225 of another seminal paper [24] by Euler which was published in 1776.

[3]This may be seen from our previous discussion of a system of forces and moments acting on a rigid body.

We now recall the developments of Section 8 of Chapter 8, where we introduced the inertia tensor:

$$\begin{aligned}
\mathbf{H} = \; & (I_{xx}\omega_x + I_{xy}\omega_y + I_{xz}\omega_z)\,\mathbf{e}_x \\
& + (I_{xy}\omega_x + I_{yy}\omega_y + I_{yz}\omega_z)\,\mathbf{e}_y \\
& + (I_{xz}\omega_x + I_{yz}\omega_y + I_{zz}\omega_z)\,\mathbf{e}_z\,.
\end{aligned}$$

Here, $I_{xx}, \ldots, I_{zz}$ are the components of the inertia tensor of the rigid body relative to the center of mass and the angular velocity vector of the rigid body is $\boldsymbol{\omega} = \omega_x\mathbf{e}_x + \omega_y\mathbf{e}_y + \omega_z\mathbf{e}_z$. In the balance laws, the derivative of $\mathbf{H}$ is required. This is obtained by differentiating the expression for $\mathbf{H}$ above. The resulting expression is conveniently written in the form

$$\dot{\mathbf{H}} = \breve{\mathbf{H}} + \boldsymbol{\omega} \times \mathbf{H}\,,$$

where $\breve{\mathbf{H}}$ is the corotational rate of $\mathbf{H}$. This is the time derivative of $\mathbf{H}$ which is obtained while keeping $\mathbf{e}_x$, $\mathbf{e}_y$, and $\mathbf{e}_z$ fixed:

$$\begin{aligned}
\breve{\mathbf{H}} = \; & (I_{xx}\dot{\omega}_x + I_{xy}\dot{\omega}_y + I_{xz}\dot{\omega}_z)\,\mathbf{e}_x \\
& + (I_{xy}\dot{\omega}_x + I_{yy}\dot{\omega}_y + I_{yz}\dot{\omega}_z)\,\mathbf{e}_y \\
& + (I_{xz}\dot{\omega}_x + I_{yz}\dot{\omega}_y + I_{zz}\dot{\omega}_z)\,\mathbf{e}_z\,.
\end{aligned}$$

Clearly, the resulting balance of angular momentum $\mathbf{M} = \dot{\mathbf{H}}$ gives rise to a complex set of equations.

9.1.3   The Fixed-Axis of Rotation Case

Simplifying the aforementioned results to the fixed-axis of rotation case:

$$\mathbf{e}_x = \cos(\theta)\mathbf{E}_x + \sin(\theta)\mathbf{E}_y\,, \quad \mathbf{e}_y = \cos(\theta)\mathbf{E}_y - \sin(\theta)\mathbf{E}_x\,, \quad \mathbf{e}_z = \mathbf{E}_z\,,$$

$$\dot{\mathbf{e}}_x = \dot{\theta}\mathbf{e}_y\,, \quad \dot{\mathbf{e}}_y = -\dot{\theta}\mathbf{e}_x\,, \quad \boldsymbol{\omega} = \dot{\theta}\mathbf{E}_z = \omega\mathbf{E}_z\,.$$

For this case,

$$\begin{aligned}
\mathbf{H} &= I_{xz}\omega\mathbf{e}_x + I_{yz}\omega\mathbf{e}_y + I_{zz}\omega\mathbf{E}_z\,, \\
\dot{\mathbf{H}} &= \left(I_{xz}\dot{\omega} - I_{yz}\omega^2\right)\mathbf{e}_x + \left(I_{yz}\dot{\omega} + I_{xz}\omega^2\right)\mathbf{e}_y + I_{zz}\dot{\omega}\mathbf{E}_z\,.
\end{aligned}$$

The balance laws for the fixed-axis of rotation case can be written as

$$\begin{aligned}
\mathbf{F} &= m\dot{\mathbf{v}}\,, \\
\mathbf{M} &= \left(I_{xz}\dot{\omega} - I_{yz}\omega^2\right)\mathbf{e}_x + \left(I_{yz}\dot{\omega} + I_{xz}\omega^2\right)\mathbf{e}_y + I_{zz}\dot{\omega}\mathbf{E}_z\,.
\end{aligned}$$

The first three of these equations give the motion of the center of mass and any reaction forces acting on the body. The fourth and fifth equations ($\mathbf{M}\cdot\mathbf{e}_x = \dot{\mathbf{H}}\cdot\mathbf{e}_x$ and $\mathbf{M}\cdot\mathbf{e}_y = \dot{\mathbf{H}}\cdot\mathbf{e}_y$) give the reaction moment $\mathbf{M}_c$ which ensures that the rotation of the rigid body is about the $\mathbf{E}_z$ axis. Last, but not least, the sixth equation ($\mathbf{M}\cdot\mathbf{E}_z = \dot{\mathbf{H}}\cdot\mathbf{E}_z$) gives a differential equation for $\theta(t)$.

9.1.4   The Four Steps

In solving problems, we will follow the four steps used earlier for particles. There are some modifications:

1. Pick an origin, a coordinate system, and a corotational basis, and then establish expressions for $\mathbf{H}$ (or $\mathbf{H}_O$), $\bar{\mathbf{x}}$, $\bar{\mathbf{v}}$, and $\bar{\mathbf{a}}$.

2. Draw a free-body diagram showing the external forces $\mathbf{F}_i$ and moments $\mathbf{M}_p$.

3. Write out the six equations $\mathbf{F} = m\bar{\mathbf{a}}$ and $\mathbf{M} = \dot{\mathbf{H}}$ (or $\mathbf{M}_O = \dot{\mathbf{H}}_O$).

4. Perform the analysis.

These steps will guide you through most problems. We emphasize once more that the free-body diagram is only used as an aid to checking one's solution.

It is a common beginner's mistake to use the balance of angular momentum about a point, say A, that is neither the center of mass C nor fixed ($\mathbf{v}_A \neq \mathbf{0}$). If one does this, then it is important to note that $\mathbf{M}_A \neq \dot{\mathbf{H}}_A$, rather $\mathbf{M}_A = \dot{\mathbf{H}}_A + \mathbf{v}_A \times \mathbf{G}$.

9.2   Work-Energy Theorem and Energy Conservation

Here, we first show the Koenig decomposition for the kinetic energy of a rigid body:

$$T = \frac{1}{2}m\bar{\mathbf{v}} \cdot \bar{\mathbf{v}} + \frac{1}{2}\mathbf{H} \cdot \boldsymbol{\omega}.$$

This is then followed by a development of the work-energy theorem for a rigid body:

$$\frac{dT}{dt} = \mathbf{F} \cdot \bar{\mathbf{v}} + \mathbf{M} \cdot \boldsymbol{\omega} = \sum_{i=1}^{K} \mathbf{F}_i \cdot \mathbf{v}_i + \mathbf{M}_p \cdot \boldsymbol{\omega}.$$

As in particles and systems of particles, this theorem can be used to establish conservation of the total energy of a rigid body during a motion.

9.2.1 Koenig's Decomposition

We begin with Koenig's[4] decomposition of the kinetic energy of a rigid body.[5] By definition, the kinetic energy T of a rigid body is

$$T = \frac{1}{2} \int_{\mathcal{R}} \mathbf{v} \cdot \mathbf{v} \rho dv \,.$$

We next recall that the velocity vector of any material point X of the rigid body has the representation

$$\mathbf{v} = \bar{\mathbf{v}} + \boldsymbol{\omega} \times \boldsymbol{\pi} \,,$$

where the relative position vector $\boldsymbol{\pi}$ and the angular velocity vector $\boldsymbol{\omega}$ are

$$\boldsymbol{\pi} = \mathbf{x} - \bar{\mathbf{x}} \,, \quad \boldsymbol{\omega} = \omega_x \mathbf{e}_x + \omega_y \mathbf{e}_y + \omega_z \mathbf{e}_z \,.$$

Substituting for $\mathbf{v}$ in the expression for T and expanding, we find that

$$T = \frac{1}{2} \int_{\mathcal{R}} \left(\bar{\mathbf{v}} \cdot \bar{\mathbf{v}} + 2\bar{\mathbf{v}} \cdot (\boldsymbol{\omega} \times \boldsymbol{\pi}) + (\boldsymbol{\omega} \times \boldsymbol{\pi}) \cdot (\boldsymbol{\omega} \times \boldsymbol{\pi}) \right) \rho dv \,.$$

However,

$$\frac{1}{2} \int_{\mathcal{R}} \bar{\mathbf{v}} \cdot \bar{\mathbf{v}} \rho dv \;=\; \frac{\bar{\mathbf{v}} \cdot \bar{\mathbf{v}}}{2} \int_{\mathcal{R}} \rho dv = \frac{1}{2} m \bar{\mathbf{v}} \cdot \bar{\mathbf{v}} \,,$$

$$\int_{\mathcal{R}} \bar{\mathbf{v}} \cdot (\boldsymbol{\omega} \times \boldsymbol{\pi}) \rho dv \;=\; \bar{\mathbf{v}} \cdot \left(\boldsymbol{\omega} \times \int_{\mathcal{R}} \boldsymbol{\pi} \rho dv \right) = \bar{\mathbf{v}} \cdot (\boldsymbol{\omega} \times \mathbf{0}) = 0 \,.$$

Consequently,

$$T = \frac{1}{2} m \bar{\mathbf{v}} \cdot \bar{\mathbf{v}} + \frac{1}{2} \int_{\mathcal{R}} (\boldsymbol{\omega} \times \boldsymbol{\pi}) \cdot (\boldsymbol{\omega} \times \boldsymbol{\pi}) \rho dv \,.$$

We can simplify the second expression of the right-hand side of this equation using a vector identity:

$$(\boldsymbol{\omega} \times \boldsymbol{\pi}) \cdot (\boldsymbol{\omega} \times \boldsymbol{\pi}) = ((\boldsymbol{\pi} \cdot \boldsymbol{\pi})\boldsymbol{\omega} - (\boldsymbol{\pi} \cdot \boldsymbol{\omega})\boldsymbol{\pi}) \cdot \boldsymbol{\omega} \,.$$

Substituting this result into the expression for T and rearranging, we find that

$$T = \frac{1}{2} m \bar{\mathbf{v}} \cdot \bar{\mathbf{v}} + \frac{1}{2} \int_{\mathcal{R}} ((\boldsymbol{\pi} \cdot \boldsymbol{\pi})\boldsymbol{\omega} - (\boldsymbol{\pi} \cdot \boldsymbol{\omega})\boldsymbol{\pi}) \rho dv \cdot \boldsymbol{\omega} \,.$$

Recall, from Section 8 of Chapter 8, that the angular momentum relative to the center of mass C of the rigid body is

$$\mathbf{H} = \int_{\mathcal{R}} \boldsymbol{\pi} \times (\boldsymbol{\omega} \times \boldsymbol{\pi}) \rho dv = \int_{\mathcal{R}} ((\boldsymbol{\pi} \cdot \boldsymbol{\pi})\boldsymbol{\omega} - (\boldsymbol{\pi} \cdot \boldsymbol{\omega})\boldsymbol{\pi}) \, \rho dv \,.$$

---

[4]Johann Samuel Koenig (1712–1757) was a German mathematican and philosopher. He was also a contemporary of Euler.

[5]Our proof of Koenig's decomposition follows Casey [11, 14].

Hence, we obtain the Koenig decomposition:

$$T = \frac{1}{2}m\bar{\mathbf{v}} \cdot \bar{\mathbf{v}} + \frac{1}{2}\mathbf{H} \cdot \boldsymbol{\omega} \, .$$

In words, the kinetic energy of a rigid body can be decomposed into the sum of the rotational kinetic energy and the (translational) kinetic energy of the center of mass.

9.2.2   The Work-Energy Theorem

To establish a work-energy theorem for a rigid body, we start by differentiating T:

$$\dot{T} = \frac{1}{2}m\dot{\bar{\mathbf{v}}} \cdot \bar{\mathbf{v}} + \frac{1}{2}m\bar{\mathbf{v}} \cdot \dot{\bar{\mathbf{v}}} + \frac{1}{2}\dot{\mathbf{H}} \cdot \boldsymbol{\omega} + \frac{1}{2}\mathbf{H} \cdot \dot{\boldsymbol{\omega}} \, .$$

To proceed further, we need to show that $\dot{\mathbf{H}} \cdot \boldsymbol{\omega} = \mathbf{H} \cdot \dot{\boldsymbol{\omega}}$. In the preceding pages all of these terms except for one are recorded. The only missing term is

$$
\begin{aligned}
\boldsymbol{\alpha} = \dot{\boldsymbol{\omega}} \;\; &= \;\; \frac{d}{dt}\left(\omega_x \mathbf{e}_x + \omega_y \mathbf{e}_y + \omega_z \mathbf{e}_z\right) \\
&= \;\; \dot{\omega}_x \mathbf{e}_x + \dot{\omega}_y \mathbf{e}_y + \dot{\omega}_z \mathbf{e}_z + \omega_x \dot{\mathbf{e}}_x + \omega_y \dot{\mathbf{e}}_y + \omega_z \dot{\mathbf{e}}_z \\
&= \;\; \dot{\omega}_x \mathbf{e}_x + \dot{\omega}_y \mathbf{e}_y + \dot{\omega}_z \mathbf{e}_z + \boldsymbol{\omega} \times \left(\omega_x \mathbf{e}_x + \omega_y \mathbf{e}_y + \omega_z \mathbf{e}_z\right) \\
&= \;\; \dot{\omega}_x \mathbf{e}_x + \dot{\omega}_y \mathbf{e}_y + \dot{\omega}_z \mathbf{e}_z \, .
\end{aligned}
$$

Another direct calculation shows that

$$
\begin{aligned}
\mathbf{H} \cdot \dot{\boldsymbol{\omega}} \;\; &= \;\; \left(I_{xx}\omega_x + I_{xy}\omega_y + I_{xz}\omega_z\right)\dot{\omega}_x \\
&\quad + \left(I_{xy}\omega_x + I_{yy}\omega_y + I_{yz}\omega_z\right)\dot{\omega}_y \\
&\quad + \left(I_{xz}\omega_x + I_{yz}\omega_y + I_{zz}\omega_z\right)\dot{\omega}_z \, .
\end{aligned}
$$

Comparing this to the corresponding expression for $\dot{\mathbf{H}} \cdot \boldsymbol{\omega}$, we find that they are equal. Consequently,

$$\dot{T} = \frac{1}{2}m\dot{\bar{\mathbf{v}}} \cdot \bar{\mathbf{v}} + \frac{1}{2}m\bar{\mathbf{v}} \cdot \dot{\bar{\mathbf{v}}} + \frac{1}{2}\dot{\mathbf{H}} \cdot \boldsymbol{\omega} + \frac{1}{2}\dot{\mathbf{H}} \cdot \boldsymbol{\omega} \, .$$

This result implies that

$$\dot{T} = m\dot{\bar{\mathbf{v}}} \cdot \bar{\mathbf{v}} + \dot{\mathbf{H}} \cdot \boldsymbol{\omega} \, .$$

Invoking the balance of linear momentum and the balance of angular momentum, we obtain the work-energy theorem:

$$\dot{T} = \mathbf{F} \cdot \bar{\mathbf{v}} + \mathbf{M} \cdot \boldsymbol{\omega} \, .$$

You should notice how this is a natural extension of the work-energy theorem for a single particle.

9.2.3   An Alternative Form of the Work-Energy Theorem

In applications, it is convenient to use an equivalent form of the work-energy theorem. This can be obtained by substituting for the moments and forces discussed at the beginning of Section 1:

$$\mathbf{F} = \sum_{i=1}^{K} \mathbf{F}_i, \quad \mathbf{M} = \mathbf{M}_p + \sum_{i=1}^{K} (\mathbf{x}_i - \bar{\mathbf{x}}) \times \mathbf{F}_i.$$

Hence, the mechanical power of the resultant forces and moments can be written as

$$\mathbf{F} \cdot \bar{\mathbf{v}} + \mathbf{M} \cdot \boldsymbol{\omega} = \left(\sum_{i=1}^{K} \mathbf{F}_i \right) \cdot \bar{\mathbf{v}} + \mathbf{M}_p \cdot \boldsymbol{\omega} + \left(\sum_{i=1}^{K} (\mathbf{x}_i - \bar{\mathbf{x}}) \times \mathbf{F}_i \right) \cdot \boldsymbol{\omega}.$$

With some minor manipulations involving the identity $\mathbf{a} \cdot (\mathbf{b} \times \mathbf{c}) = \mathbf{c} \cdot (\mathbf{a} \times \mathbf{b})$, and noting that $\mathbf{v}_i = \bar{\mathbf{v}} + \boldsymbol{\omega} \times (\mathbf{x}_i - \bar{\mathbf{x}})$, we find that

$$\mathbf{F} \cdot \bar{\mathbf{v}} + \mathbf{M} \cdot \boldsymbol{\omega} = \sum_{i=1}^{K} \mathbf{F}_i \cdot \mathbf{v}_i + \mathbf{M}_p \cdot \boldsymbol{\omega}.$$

In conclusion, we have an alternative form of the work-energy theorem that proves to be far easier to use in applications:

$$\dot{T} = \sum_{i=1}^{K} \mathbf{F}_i \cdot \mathbf{v}_i + \mathbf{M}_p \cdot \boldsymbol{\omega}.$$

We shall present examples shortly involving this theorem to illustrate how it is used to establish conservation of energy results.

It is crucial to note from the work-energy theorem what the mechanical powers of a force $\mathbf{P}$ acting at a material point X whose position vector is $\mathbf{x}$ and a moment $\mathbf{L}$ are

Mechanical power of a force $\mathbf{P}$ acting at $\mathbf{x}$: $\mathbf{P} \cdot \dot{\mathbf{x}}$,
Mechanical power of a moment $\mathbf{L}$: $\mathbf{L} \cdot \boldsymbol{\omega}$.

These expressions can be easily used to determine whether a force or moment is workless, and thus does not contribute to the change of kinetic energy of the rigid body during a motion.

9.3   Purely Translational Motion of a Rigid Body

A rigid body performing a purely translational motion is arguably the simplest class of problems associated with these bodies. For these problems, the angular velocity $\boldsymbol{\omega}$ and acceleration $\boldsymbol{\alpha}$ vectors are $\mathbf{0}$. The velocity and acceleration of any material point of the rigid body are none other than those for the center of mass C. Recalling the expression for the angular momentum $\mathbf{H}$,

$$
\begin{aligned}
\mathbf{H} \ = \ & (I_{xx}\omega_x + I_{xy}\omega_y + I_{xz}\omega_z)\,\mathbf{e}_x \\
& + (I_{xy}\omega_x + I_{yy}\omega_y + I_{yz}\omega_z)\,\mathbf{e}_y \\
& + (I_{xz}\omega_x + I_{yz}\omega_y + I_{zz}\omega_z)\,\mathbf{e}_z\,,
\end{aligned}
$$

one also finds that, for these problems, $\mathbf{H}$ and $\dot{\mathbf{H}}$ are $\mathbf{0}$.

For purely translational problems, the balance laws are simply

$$
\mathbf{F} = m\dot{\mathbf{v}}\,, \quad \mathbf{M} = \mathbf{0}\,.
$$

These give 6 equations to solve for constraint forces and moments and the motion of the center of mass of the rigid body. In addition, the work-energy theorem is simply

$$
\dot{T} = \mathbf{F} \cdot \bar{\mathbf{v}}\,.
$$

9.3.1 The Overturning Cart

We now consider an example. As shown in Figure 9.2, a cart of mass m, height $2a$, width $2b$, and depth $2c$ is being driven by a force $\mathbf{P} = P\mathbf{E}_x$. This force is applied to a point on one of its sides. The cart is free to move on a smooth horizontal track. We wish to determine the restrictions on P such that the cart will not topple. In addition, we will prove that the total energy of the cart is conserved.

You should notice how the solution of this problem follows the four steps we discussed earlier.

Kinematics

We start with the kinematics, and choose a Cartesian coordinate system to describe $\bar{\mathbf{x}}$:

$$
\bar{\mathbf{x}} = x\mathbf{E}_x + y_0\mathbf{E}_y + z_0\mathbf{E}_z\,,
$$

where y_0 and z_0 are constant because we will consider only the case where all four wheels of the cart remain in contact with the track. Differentiating this expression gives the velocity and acceleration vectors of the center of mass. For the present problem, one does not need to establish an expression for $\mathbf{H}$. Further, we find no need to explicitly mention the existence of a corotational basis $\{\mathbf{e}_x, \mathbf{e}_y, \mathbf{e}_z\}$ since we can choose $\mathbf{e}_x = \mathbf{E}_x$, $\mathbf{e}_y = \mathbf{E}_y$, and $\mathbf{e}_z = \mathbf{E}_z$.

Forces and Moments

We next consider the free-body diagram (shown in Figure 9.3). Notice that there are four reaction forces, one on each of the four wheels: $\mathbf{N}_i$ acts on

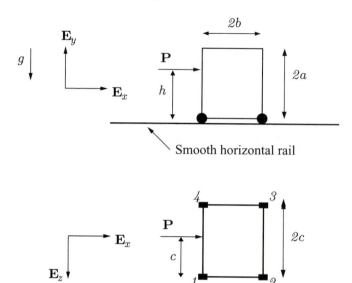

FIGURE 9.2. A cart moving on a smooth horizontal track

the wheel numbered i in Figure 9.2, where $i = 1,\ 2,\ 3$ or 4. These forces have components in the $\mathbf{E}_y$ and $\mathbf{E}_z$ directions.[6]

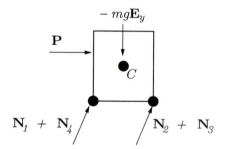

FIGURE 9.3. Free-body diagram of the cart

The resultant force acting on the system is

$$
\mathbf{F} \;=\; P\mathbf{E}_x - mg\mathbf{E}_y + \left(N_{1y} + N_{2y} + N_{3y} + N_{4y}\right)\mathbf{E}_y
$$
$$
+ \left(N_{1z} + N_{2z} + N_{3z} + N_{4z}\right)\mathbf{E}_z .
$$

[6]The components of the normal forces in the $\mathbf{E}_z$ direction ensure that the cart will not start to rotate about the $\mathbf{E}_y$ axis.

The resultant moment can be calculated by taking the moments of these forces about the center of mass:

$$
\begin{aligned}
\mathbf{M} \;=\; & ((h-a)\mathbf{E}_y - b\mathbf{E}_x) \times P\mathbf{E}_x + \mathbf{0} \times -mg\mathbf{E}_y \\
& + (-a\mathbf{E}_y - b\mathbf{E}_x + c\mathbf{E}_z) \times (N_{1y}\mathbf{E}_y + N_{1z}\mathbf{E}_z) \\
& + (-a\mathbf{E}_y + b\mathbf{E}_x + c\mathbf{E}_z) \times (N_{2y}\mathbf{E}_y + N_{2z}\mathbf{E}_z) \\
& + (-a\mathbf{E}_y + b\mathbf{E}_x - c\mathbf{E}_z) \times (N_{3y}\mathbf{E}_y + N_{3z}\mathbf{E}_z) \\
& + (-a\mathbf{E}_y - b\mathbf{E}_x - c\mathbf{E}_z) \times (N_{4y}\mathbf{E}_y + N_{4z}\mathbf{E}_z) \,.
\end{aligned}
$$

With some work, the expression for $\mathbf{M}$ can be simplified:

$$
\begin{aligned}
\mathbf{M} \;=\; & (a-h)P\mathbf{E}_z + c(-N_{1y} - N_{2y} + N_{3y} + N_{4y})\mathbf{E}_x \\
& - a(N_{1z} + N_{2z} + N_{3z} + N_{4z})\mathbf{E}_x + b(N_{1z} - N_{2z} - N_{3z} + N_{4z})\mathbf{E}_y \\
& + b(-N_{1y} + N_{2y} + N_{3y} - N_{4y})\mathbf{E}_z \,.
\end{aligned}
$$

Balance Laws

We now invoke the balance laws, and take their components with respect to $\mathbf{E}_x$, $\mathbf{E}_y$, and $\mathbf{E}_z$ to obtain the 6 equations

$$
\begin{aligned}
P &= m\ddot{x}\,, \\
-mg + N_{1y} + N_{2y} + N_{3y} + N_{4y} &= 0\,, \\
N_{1z} + N_{2z} + N_{3z} + N_{4z} &= 0\,, \\
-c(N_{1y} + N_{2y} - N_{3y} - N_{4y}) &= a(N_{1z} + N_{2z} + N_{3z} + N_{4z})\,, \\
N_{1z} - N_{2z} - N_{3z} + N_{4z} &= 0\,, \\
(h-a)P &= b(-N_{1y} + N_{2y} + N_{3y} - N_{4y})\,.
\end{aligned}
$$

There are 9 unknowns, $x(t)$, N_{1y}, N_{2y}, N_{3y}, N_{4y}, N_{1z}, N_{2z}, N_{3z}, and N_{4z}. It follows that the system of equations is indeterminate. We will address this matter shortly.

Analysis

First, let's determine the motion of the system. From the 6 equations given above, we see that

$$
\ddot{x} = \frac{P}{m}\,.
$$

Subject to the initial conditions at $t = 0$ that $\bar{\mathbf{x}}(0) = a\mathbf{E}_z$ and $\bar{\mathbf{v}}(0) = v_0\mathbf{E}_x$, we can integrate this equation to find that the motion of the center of mass is

$$
\bar{\mathbf{x}} = \left(v_0 t + \frac{Pt^2}{2m}\right)\mathbf{E}_x + a\mathbf{E}_z\,.
$$

In words, the cart's center of mass is accelerated in the direction of the applied force $P\mathbf{E}_x$: a result that shouldn't surprise you.

Next, we seek to determine the toppling force. From the 6 equations above, we obtain 5 equations for the 8 unknown reaction forces. As mentioned earlier, this is an indeterminate system of equations. To proceed, we need to make 3 additional assumptions.[7] These assumptions relate the forces on the individual wheels:

$$N_{1y} = N_{4y}, \quad N_{2y} = N_{3y}, \quad N_{1z} = -N_{3z}.$$

That is, the vertical forces on the rear wheels are identical, the vertical forces on the front wheels are identical, and the horizontal force on one of the rear wheels is equal and opposite to that on one of the front wheels. Solving for the reaction forces we find that

$$N_{1z} = N_{2z} = N_{3z} = N_{4z} \quad = \quad 0,$$

$$N_{1y} = N_{4y} \quad = \quad \frac{1}{4}\left(mg - \frac{P}{b}(h-a)\right),$$

$$N_{2y} = N_{3y} \quad = \quad \frac{1}{4}\left(mg + \frac{P}{b}(h-a)\right).$$

To determine the allowable range of P, we set the $\mathbf{E}_y$ components of the reaction forces to zero. Consequently, the front wheels will lose contact, provided that $N_{2y} = N_{3y} < 0$:

$$P < -\frac{mgb}{h-a}.$$

Similarly, the rear wheels loose contact when

$$P > \frac{mgb}{h-a}.$$

For a given cart and h, one can now easily determine the allowable range of P. We leave this as an exercise. In the course of this exercise, you will notice that if a cart is tall (i.e., $a \gg b$) then the toppling force P is smaller than if the cart were stout (i.e., $a \ll b$).

We now address energy conservation for this problem. Starting from the work-energy theorem, we find that

$$\dot{T} = \sum_{i=1}^{K} \mathbf{F}_i \cdot \bar{\mathbf{v}}$$

$$= (-mg + N_{1y} + N_{2y} + N_{3y} + N_{4y})\mathbf{E}_y \cdot \bar{\mathbf{v}}$$

$$+ (N_{1z} + N_{2z} + N_{3z} + N_{4z})\mathbf{E}_z \cdot \bar{\mathbf{v}} + P\mathbf{E}_x \cdot \bar{\mathbf{v}}.$$

---

[7]Of course, the correct approach here would be to model each of the four wheels attached to the cart as individual rigid bodies. Then, instead of modeling this system as a single rigid body, one has a system of five rigid bodies. Unfortunately, one has a similar indeterminacy in this model also. In vehicle system dynamics, an area that is primarily concerned with modeling automobiles using interconnected rigid bodies, this issue is usually not seen because one incorporates suspension models (see Gillespie [27]).

However, the normal forces and gravity have no mechanical power in this problem and P is constant:

$$\dot{T} = P\mathbf{E}_x \cdot \mathbf{v} = \frac{d}{dt}\left(P\mathbf{E}_x \cdot \mathbf{x}\right).$$

Hence, the total energy E of this system is conserved:

$$\frac{d}{dt}\left(E = T - P\mathbf{E}_x \cdot \bar{\mathbf{x}} = \frac{1}{2}m\dot{x}^2 - Px\right) = 0.$$

Notice that for this problem, $T = 0.5m\bar{\mathbf{v}} \cdot \bar{\mathbf{v}}$ is the kinetic energy of the center of mass.

9.4 A Rigid Body with a Fixed Point

In the class of problem considered in this section, a rigid body is attached at one of its material points by a pin-joint to a fixed point (cf. Figure 9.4). Here, we take this fixed point to be the origin O of our coordinate system. At the pin-joint there is a reaction force $\mathbf{R}$ and a reaction moment $\mathbf{M}_c$. These ensure that the point of attachment remains fixed and axis of rotation of the rigid body remains fixed, respectively.

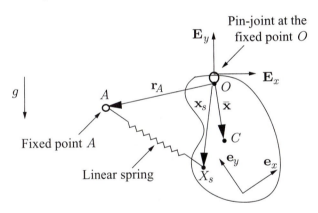

FIGURE 9.4. A representative example of a rigid body rotating about a fixed point O

The example shown in Figure 9.4 is representative. We assume that a gravitational force $-mg\mathbf{E}_y$ acts on the body, in addition to a spring force:

$$\mathbf{F}_s = -K(\|\mathbf{x}_s - \mathbf{r}_A\| - L)\frac{\mathbf{x}_s - \mathbf{r}_A}{\|\mathbf{x}_s - \mathbf{r}_A\|}.$$

One end of the spring is attached to the material point X_s of the rigid body whose position vector is $\mathbf{x}_S$. The other end is attached to a fixed point whose position vector is $\mathbf{r}_A$.

Restricting attention to planar motions, we seek to determine the differential equation governing the motion of this rigid body, the reaction forces, and the reaction moments. We also prove that the total energy of this rigid body is conserved.

9.4.1 Kinematics

Since the motion is constrained to be planar, $\boldsymbol{\omega} = \dot{\theta}\mathbf{E}_z = \omega\mathbf{E}_z$ and $\boldsymbol{\alpha} = \ddot{\theta}\mathbf{E}_z$. The corotational basis is defined in the usual manner:

$$\mathbf{e}_x = \cos(\theta)\mathbf{E}_x + \sin(\theta)\mathbf{E}_y \,, \quad \mathbf{e}_y = -\sin(\theta)\mathbf{E}_x + \cos(\theta)\mathbf{E}_y \,, \quad \mathbf{e}_z = \mathbf{E}_z \,.$$

We also recall the relations

$$\dot{\mathbf{e}}_x = \dot{\theta}\mathbf{e}_y \,, \quad \dot{\mathbf{e}}_y = -\dot{\theta}\mathbf{e}_x \,, \quad \dot{\mathbf{e}}_z = \mathbf{0} \,.$$

Taking the fixed point O as the origin, we denote the position vector of the center of mass of the body by

$$\bar{\mathbf{x}} = x\mathbf{e}_x + y\mathbf{e}_y + z\mathbf{e}_z \,.$$

Here, x, y, and z are constants. Differentiating this expression, we find that

$$\bar{\mathbf{v}} = \dot{\theta}(x\mathbf{e}_y - y\mathbf{e}_x) \,, \quad \bar{\mathbf{a}} = \ddot{\theta}(x\mathbf{e}_y - y\mathbf{e}_x) - \dot{\theta}^2(x\mathbf{e}_x + y\mathbf{e}_y) \,.$$

We also define the position vectors of the spring's attachment points:

$$\mathbf{x}_s = x_s\mathbf{e}_x + y_s\mathbf{e}_y + z_s\mathbf{e}_z \,, \quad \mathbf{r}_A = X_A\mathbf{E}_x + Y_A\mathbf{E}_y + Z_A\mathbf{E}_z \,,$$

where all of the displayed coordinates are constant. To keep the development clear, we will avoid explicit use of these representations for $\mathbf{x}_s$ and $\mathbf{r}_A$.

Next, we address the angular momenta of the rigid body. Since the axis of rotation is fixed, we recall from Section 1 that

$$\begin{aligned}
\mathbf{H} &= I_{xz}\omega\mathbf{e}_x + I_{yz}\omega\mathbf{e}_y + I_{zz}\omega\mathbf{E}_z \,, \\
\dot{\mathbf{H}} &= \left(I_{xz}\dot{\omega} - I_{yz}\omega^2\right)\mathbf{e}_x + \left(I_{yz}\dot{\omega} + I_{xz}\omega^2\right)\mathbf{e}_y + I_{zz}\dot{\omega}\mathbf{E}_z \,.
\end{aligned}$$

It is convenient in this problem to determine $\mathbf{H}_O$. To this end, we recall that

$$\mathbf{H}_O = \mathbf{H} + \bar{\mathbf{x}} \times \mathbf{G} \,.$$

Substituting for the problem of interest, we obtain

$$\mathbf{H}_O = I_{xz}\omega\mathbf{e}_x + I_{yz}\omega\mathbf{e}_y + I_{zz}\omega\mathbf{E}_z + (x\mathbf{e}_x + y\mathbf{e}_y + z\mathbf{e}_z) \times m\dot{\theta}(x\mathbf{e}_y - y\mathbf{e}_x) \,.$$

With some rearranging, we find that

$$\mathbf{H}_O = (I_{xz} - mxz)\,\omega\mathbf{e}_x + (I_{yz} - myz)\,\omega\mathbf{e}_y + \left(I_{zz} + m\left(x^2 + y^2\right)\right)\omega\mathbf{E}_z \,.$$

These results are identical to those that would have been obtained had one used the parallel-axis theorem to determine the moment of inertia tensor of the body about point O.

For future purposes, we also determine the kinetic energy T of the rigid body. We start from the Koenig decomposition and substitute for the various kinematical quantities to find that

$$
\begin{aligned}
T &= \frac{1}{2}m\bar{\mathbf{v}} \cdot \bar{\mathbf{v}} + \frac{1}{2}\mathbf{H} \cdot \boldsymbol{\omega} \\
&= \frac{1}{2}\left(I_{zz} + m\left(x^2 + y^2\right)\right)\dot{\theta}^2.
\end{aligned}
$$

Notice that $T = \frac{1}{2}\mathbf{H}_O \cdot \boldsymbol{\omega}$ for this problem.

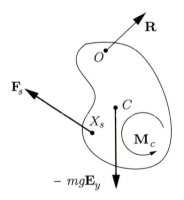

FIGURE 9.5. Free-body diagram

9.4.2   Forces and Moments

The free-body diagram for the body is shown in Figure 9.5. Notice that there is a reaction force $\mathbf{R}$ at O:

$$
\mathbf{R} = R_x\mathbf{E}_x + R_y\mathbf{E}_y + R_z\mathbf{E}_z.
$$

In addition, there is a reaction moment $\mathbf{M}_c$ acting on the body:

$$
\mathbf{M}_c = M_{cx}\mathbf{e}_x + M_{cy}\mathbf{e}_y.
$$

This moment ensures that the rotation of the body is constrained to being about the $\mathbf{E}_z$ axis. This is normally not mentioned in engineering dynamics texts. We shall shortly see why it is needed. The resultant force and moment on the body are

$$
\mathbf{F} = -mg\mathbf{E}_y + \mathbf{F}_s + \mathbf{R}, \quad \mathbf{M}_O = \mathbf{M}_c - \bar{\mathbf{x}} \times mg\mathbf{E}_y + \mathbf{x}_s \times \mathbf{F}_s.
$$

9.4.3 Balance Laws

We now turn to the balances of linear and angular momenta for the rigid body. Due to the unknown force $\mathbf{R}$, it is easiest to use the balance of angular momentum about O. The balance laws are

$$\mathbf{F} = m\bar{\mathbf{a}}, \quad \mathbf{M}_O = \dot{\mathbf{H}}_O.$$

Substituting for the resultant forces and moments, we find that

$$\mathbf{F}_s - mg\mathbf{E}_y + \mathbf{R} = m\bar{\mathbf{a}}, \quad \mathbf{M}_c - \bar{\mathbf{x}} \times mg\mathbf{E}_y + \mathbf{x}_s \times \mathbf{F}_s = \dot{\mathbf{H}}_O.$$

We have refrained from substituting for the inertias and accelerations here. These 6 equations may be solved for the 5 unknown reactions, $\mathbf{R}$ and $\mathbf{M}_c$, and they also provide a differential equation governing the motion of the rigid body.

9.4.4 Analysis

We first determine the unknown forces and moments. From the balance of linear momentum, we obtain 3 equations for the 3 unknown components of $\mathbf{R}$:

$$\begin{aligned}
\mathbf{R} &= mg\mathbf{E}_y - \mathbf{F}_s + m\bar{\mathbf{a}} \\
&= mg\mathbf{E}_y + K(\|\mathbf{x}_s - \mathbf{r}_A\| - L)\frac{\mathbf{x}_s - \mathbf{r}_A}{\|\mathbf{x}_s - \mathbf{r}_A\|} \\
&\quad + m\left(\ddot{\theta}(x\mathbf{e}_y - y\mathbf{e}_x) - \dot{\theta}^2(x\mathbf{e}_x + y\mathbf{e}_y)\right).
\end{aligned}$$

Next, we find, from the balance of angular momentum, that

$$\begin{aligned}
M_{cx} = \mathbf{M}_c \cdot \mathbf{e}_x &= \left(-\mathbf{x}_s \times \mathbf{F}_s + \bar{\mathbf{x}} \times mg\mathbf{E}_y + \dot{\mathbf{H}}_O\right) \cdot \mathbf{e}_x, \\
M_{cy} = \mathbf{M}_c \cdot \mathbf{e}_y &= \left(-\mathbf{x}_s \times \mathbf{F}_s + \bar{\mathbf{x}} \times mg\mathbf{E}_y + \dot{\mathbf{H}}_O\right) \cdot \mathbf{e}_y.
\end{aligned}$$

The expression for the spring force $\mathbf{F}_s$ and rate of change of angular momentum $\dot{\mathbf{H}}_O$ can be substituted into these expressions if desired. Notice that if the reaction moment were omitted, then the $\mathbf{e}_x$ and $\mathbf{e}_y$ components of the balance of angular momentum wouldn't be satisfied. As a result the body couldn't move as one had assumed when setting up the kinematics of the problem.[8]

The motion of the rigid body can be found from the sole remaining equation. This equation is the $\mathbf{e}_z$ component of the balance of angular momentum:

$$(\mathbf{x}_s \times \mathbf{F}_s - \bar{\mathbf{x}} \times mg\mathbf{E}_y) \cdot \mathbf{E}_z = \dot{\mathbf{H}}_O \cdot \mathbf{E}_z.$$

---

[8]The error in neglecting $\mathbf{M}_c$ is equivalent to ignoring the normal force acting on a particle that is assumed to move on a surface.

Substituting for the forces and momentum, we find that

$$\left(I_{zz} + m\left(x^2 + y^2\right)\right)\ddot{\theta} = -mg\left(x\cos(\theta) - y\sin(\theta)\right)$$
$$+ K(\|\mathbf{x}_s - \mathbf{r}_A\| - L)\frac{(\mathbf{x}_s \times \mathbf{r}_A) \cdot \mathbf{E}_z}{\|\mathbf{x}_s - \mathbf{r}_A\|}.$$

Given the initial conditions $\theta(t_0) = \theta_0$ and $\dot{\theta}(t_0) = \omega_0$, the solution of this equation determines the motion of the body.[9]
For this type of problem, one could also take the balance of angular momentum relative to the center of mass C. In this case, the reaction force $\mathbf{R}$ would make its presence known in all 6 equations. By taking the balance of angular momentum relative to the point O, this reaction force doesn't enter 3 of the 6 equations and, consequently, makes the system of 6 equations easier to solve.
Let us now prove why the total energy E of the system is conserved. We start with the (alternative form of the) work-energy theorem and substitute for the forces and moments

$$\dot{T} = \mathbf{R} \cdot \mathbf{v}_O - mg\mathbf{E}_y \cdot \bar{\mathbf{v}} + \mathbf{M}_c \cdot \boldsymbol{\omega} + \mathbf{F}_s \cdot \mathbf{v}_s.$$

An expression for the kinetic energy T for this problem was recorded earlier. Now, because O is fixed, $\mathbf{v}_O = \mathbf{0}$, so $\mathbf{R}$ has no mechanical power. Similarly, $\mathbf{M}_c$ is perpendicular to $\boldsymbol{\omega}$, so it too has no power. The spring and gravitational forces are conservative:

$$-mg\mathbf{E}_y \cdot \bar{\mathbf{v}} = -\frac{d}{dt}\left(mg\mathbf{E}_y \cdot \bar{\mathbf{x}}\right), \quad \mathbf{F}_s \cdot \mathbf{v}_s = -\frac{d}{dt}\left(\frac{K}{2}(\|\mathbf{x}_s - \mathbf{r}_A\| - L)^2\right).$$

Combining these results, we find that the total energy E of the rigid body is conserved:

$$\frac{d}{dt}\left(E = T + mg\mathbf{E}_y \cdot \bar{\mathbf{x}} + \frac{K}{2}(\|\mathbf{x}_s - \mathbf{r}_A\| - L)^2\right) = 0.$$

One uses this equation in a similar manner as with particles. For instance, given the initial conditions $\theta(t_0) = \theta_0$ and $\dot{\theta}(t_0) = \omega_0$, one can then use the conservation of E to determine $\dot{\theta}$ at a later instant of the motion when θ is given.

9.5   Rolling and Sliding Rigid Bodies

We now return to the rolling and sliding rigid bodies considered in Sections 5 and 6 of Chapter 8. We start by considering a rigid body $\mathcal{B}$ that is

---

[9]This is a nonlinear differential equation. As mentioned for a related example in the Appendix, its analytical solution can be found and expressed in terms of Jacobi's elliptic functions. The interested reader is referred to Lawden [36] or Whittaker [67] for details on these functions and how they are used in the solution of dynamics problems.

in motion atop a fixed surface $\mathcal{S}$ (see Figure 9.6). At the instantaneous point of contact $P = X_P(t)$, the outward unit normal to the surface is $\mathbf{n}$. The velocity vector of the material point $X_P(t)$ that is instantaneously in contact with the surface is denoted by $\mathbf{v}_P$. Earlier, we saw that if the rigid body is rolling on the surface, then one has the rolling condition:

$$\mathbf{v}_P = \bar{\mathbf{v}} + \boldsymbol{\omega} \times (\mathbf{r}_P - \bar{\mathbf{x}}) = \mathbf{0}.$$

If the rigid body is sliding on the surface, then one has the sliding condition:

$$\mathbf{v}_P \cdot \mathbf{n} = \bar{\mathbf{v}} \cdot \mathbf{n} + (\boldsymbol{\omega} \times (\mathbf{r}_P - \bar{\mathbf{x}})) \cdot \mathbf{n} = 0.$$

We now turn to the forces that enforce these constraints.

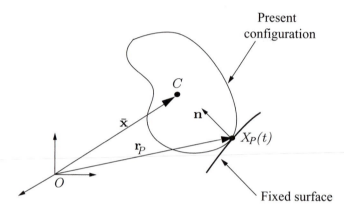

FIGURE 9.6. The geometry of contact

9.5.1 Friction

The force at the instantaneous point of contact P depends on the natures of the outer surface of the rigid body and the fixed surface. If their contact is smooth, then the reaction force at P is

$$\mathbf{F}_P = N\mathbf{n}.$$

On the other hand, if the surface is rough with coefficients of friction μ_s and μ_d, then this force is either of the static type ($\mathbf{v}_P = \mathbf{0}$),

$$\mathbf{F}_P = F_{px}\mathbf{E}_x + F_{py}\mathbf{E}_y + F_{pz}\mathbf{E}_z, \quad \text{where } \|\mathbf{F}_P - (\mathbf{F}_P \cdot \mathbf{n})\mathbf{n}\| \leq \mu_s |\mathbf{F}_P \cdot \mathbf{n}|,$$

or, if there is relative motion ($\mathbf{v}_P \neq \mathbf{0}$),

$$\mathbf{F}_P = N\mathbf{n} - \mu_d \|N\mathbf{n}\| \frac{\mathbf{v}_P}{\|\mathbf{v}_P\|}.$$

Clearly, if the contact is rough, then the rigid body can either roll or slip depending primarily on the amount of static friction available.

9.5.2   Energy Considerations

If the contact is smooth, then it should be clear that the force $\mathbf{F}_P$ is workless. Similarly, if the rigid body is rolling, then the mechanical power of $\mathbf{F}_P$ is $\mathbf{F}_P \cdot \mathbf{v}_P = 0$. It is only when the rigid body is sliding and the contact is rough that $\mathbf{F}_P$ does work:

$$\mathbf{F}_P \cdot \mathbf{v}_P = N\mathbf{n} \cdot \mathbf{v}_P - \mu_d \|N\mathbf{n}\| \frac{\mathbf{v}_P}{\|\mathbf{v}_P\|} \cdot \mathbf{v}_P = -\mu_d \|N\mathbf{n}\| \|\mathbf{v}_P\| < 0 \,.$$

Notice that the power of the force in this case is negative, so it will decrease the kinetic energy T.

The previous results imply that if the only other forces acting on a rolling rigid body are conservative (such as gravitational and spring forces), then the total energy of the rigid body will be conserved. A similar comment applies to a sliding rigid body when the contact is smooth. As a result, in most solved problems in this area, such as sliding disks and tops, and rolling disks and spheres, this energy conservation is present.[10]

9.6   Examples of Rolling and Sliding Rigid Bodies

Let us now turn to a specific example, which is shown in Figure 9.7. We consider a rigid body of mass m, whose outer surface is circular with radius R. We assume that the rigid body moves on an inclined plane under the influence of gravity. The center of mass C of the rigid body is assumed to be located at the geometric center of the circle of radius R. The contact is assumed to be rough with coefficients of friction μ_s and μ_d. It is assumed that $\mu_s > \mu_d$.

9.6.1   General Considerations

We assume that the center of mass of the rigid body is given an initial velocity $v_0 \mathbf{E}_x$, where $v_0 > 0$, at time $t = 0$, and we seek to determine the motion of the rigid body for subsequent times.

Kinematics

We assume that the motion is such that the axis of rotation is fixed: $\boldsymbol{\omega} = \dot{\theta}\mathbf{E}_z$ and $\boldsymbol{\alpha} = \ddot{\theta}\mathbf{E}_z$. The corotational basis is defined in the usual manner:

$$\mathbf{e}_x = \cos(\theta)\mathbf{E}_x + \sin(\theta)\mathbf{E}_y \,, \quad \mathbf{e}_y = -\sin(\theta)\mathbf{E}_x + \cos(\theta)\mathbf{E}_y \,, \quad \mathbf{e}_z = \mathbf{E}_z \,.$$

---

[10]There are few completely solved problems in this area (see, for examples and references, Hermans [31], Neimark and Fufaev [42], O'Reilly [45], Routh [51, 52], and Zenkov, Bloch, and Marsden [68]). It is unfortunate that the seminal work of Sergei Alekseevich Chaplygin (1869–1942) in this area has not been translated from Russian into English.

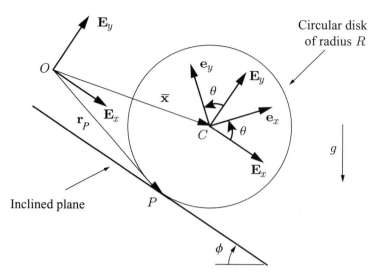

FIGURE 9.7. A rigid body on an inclined plane

We also recall the relations

$$\dot{\mathbf{e}}_x = \dot{\theta}\mathbf{e}_y, \quad \dot{\mathbf{e}}_y = -\dot{\theta}\mathbf{e}_x, \quad \dot{\mathbf{e}}_z = \mathbf{0}.$$

Taking the origin as shown in Figure 9.7, we denote the position vector of the center of mass of the body by

$$\bar{\mathbf{x}} = x\mathbf{E}_x + y\mathbf{E}_y + z_0\mathbf{E}_z.$$

Here, z_0 is a constant. Differentiating this expression, we find that

$$\bar{\mathbf{v}} = \dot{x}\mathbf{E}_x + \dot{y}\mathbf{E}_y, \quad \bar{\mathbf{a}} = \ddot{x}\mathbf{E}_x + \ddot{y}\mathbf{E}_y.$$

Next, we address the angular momentum of the rigid body. We assume that $\{\mathbf{e}_x, \mathbf{e}_y, \mathbf{e}_z\}$ are principal axes of the rigid body in its present configuration. Hence,

$$\mathbf{H} = I_{zz}\dot{\theta}\mathbf{E}_z, \quad \dot{\mathbf{H}} = I_{zz}\ddot{\theta}\mathbf{E}_z.$$

If the contact condition is such that sliding occurs, then, as the normal $\mathbf{n} = \mathbf{E}_y$,

$$\mathbf{v}_P \cdot \mathbf{E}_y = \bar{\mathbf{v}} \cdot \mathbf{E}_y + (\boldsymbol{\omega} \times (\mathbf{r}_P - \bar{\mathbf{x}})) \cdot \mathbf{E}_y = 0.$$

For the present problem, $\mathbf{r}_P - \bar{\mathbf{x}} = -R\mathbf{E}_y$, so the sliding condition implies that $\dot{y} = 0$, as expected. For sliding, one has

$$\mathbf{v}_P = \left(\dot{x} + R\dot{\theta}\right)\mathbf{E}_x = v_P\mathbf{E}_x.$$

The velocity v_P is often referred to as the slip velocity. On the other hand, if rolling occurs, then

$$\mathbf{v}_P = \bar{\mathbf{v}} + \boldsymbol{\omega} \times (\mathbf{r}_P - \bar{\mathbf{x}}) = \mathbf{0}.$$

This implies for the present problem that

$$\bar{\mathbf{v}} = \dot{x}\mathbf{E}_x = -R\dot{\theta}\mathbf{E}_x \, .$$

We discussed these results previously in Section 6 of Chapter 8.

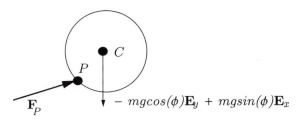

FIGURE 9.8. Free-body diagram

Forces and Moments

One of the tasks remaining is to describe $\mathbf{F}$ and $\mathbf{M}$. The free-body diagram for the system is shown in Figure 9.8. In sum,

$$
\begin{aligned}
\mathbf{F} &= -mg\cos(\phi)\mathbf{E}_y + mg\sin(\phi)\mathbf{E}_x + \mathbf{F}_P \, , \\
\mathbf{M} &= -R\mathbf{E}_y \times \mathbf{F}_P \, .
\end{aligned}
$$

If the body is rolling, then

$$\mathbf{F}_P = F_{Px}\mathbf{E}_x + F_{Py}\mathbf{E}_y + F_{Pz}\mathbf{E}_z \, ,$$

while if the body is sliding, then

$$\mathbf{F}_P = N\mathbf{E}_y - \mu_d\|N\mathbf{E}_y\|\frac{\mathbf{v}_P}{\|\mathbf{v}_P\|} = N\mathbf{E}_y - \mu_d|N|\frac{\dot{x} + R\dot{\theta}}{|\dot{x} + R\dot{\theta}|}\mathbf{E}_x \, .$$

Balance Laws

We are now in a position to examine the balance laws:

$$
\begin{aligned}
\mathbf{F}_P - mg\cos(\phi)\mathbf{E}_y + mg\sin(\phi)\mathbf{E}_x &= m\ddot{x}\mathbf{E}_x \, , \\
-R\mathbf{E}_y \times \mathbf{F}_P &= I_{zz}\ddot{\theta}\mathbf{E}_z \, .
\end{aligned}
$$

These equations are used to determine the force at P and the motion of the rigid body in what follows.

9.6.2 The Rolling Case

We first find, from the balance of linear momentum, that

$$\mathbf{F}_P = mg\cos(\phi)\mathbf{E}_y - mg\sin(\phi)\mathbf{E}_x + m\ddot{x}\mathbf{E}_x \, .$$

Substituting this result into the balance of angular momentum, and using the condition $\ddot{x} = -R\ddot{\theta}$, one obtains a differential equation for $x(t)$:

$$\frac{I_{zz} + mR^2}{R}\ddot{x} = mgR\sin(\phi)\,.$$

This equation can be easily solved to determine $x(t)$:

$$x(t) = \frac{mgR^2\sin(\phi)}{2(I_{zz} + mR^2)}t^2 + v_0 t\,.$$

In writing this solution, we assumed that $x(t = 0) = 0$. We have also tacitly assumed, in order to satisfy the rolling condition, that $\dot{\theta}(t = 0) = -\frac{v_0}{R}$. We can also find the friction and normal forces as functions of time by substituting for $x(t)$ in the expression for $\mathbf{F}_P$ given above:

$$\mathbf{F}_P = mg\cos(\phi)\mathbf{E}_y - \frac{mg\sin(\phi)I_{zz}}{I_{zz} + mR^2}\mathbf{E}_x\,.$$

This is the complete solution to the rolling case.[11]

9.6.3 The Static Friction Criterion and Rolling

To determine whether there is a transition to sliding, we need to check the magnitude of the friction force for the rolling rigid body. Here, we use the standard static friction criterion:

$$\|\mathbf{F}_{Px}\mathbf{E}_x\| \leq \mu_s\|\mathbf{F}_{Py}\mathbf{E}_y\|\,.$$

Substituting for $\mathbf{F}_P$, we find that

$$\| -\frac{mg\sin(\phi)I_{zz}}{I_{zz} + mR^2}\mathbf{E}_x\| \leq \mu_s\|mg\cos(\phi)\mathbf{E}_y\|\,,$$

which can be simplified to

$$\frac{\tan(\phi)I_{zz}}{I_{zz} + mR^2} \leq \mu_s\,.$$

If the incline is sufficiently steep, or the inertia is distributed in a certain manner, then this inequality will be violated. Its violation will start at the initial instant of the rigid body's motion. In other words, if the rigid body rolls in this problem, then it will always roll.

---

[11]It is a good exercise to show that the total energy of this rolling rigid body is conserved. If this rigid body slips, then you should also be able to show that because $\mathbf{F}_P \cdot \mathbf{v}_P < 0$ the total energy will decrease with time.

9.6.4   The Sliding Case

For a sliding rigid body, the balance laws yield

$$N\mathbf{E}_y - \mu_d|N|\frac{\dot{x} + R\dot{\theta}}{|\dot{x} + R\dot{\theta}|}\mathbf{E}_x - mg\cos(\phi)\mathbf{E}_y + mg\sin(\phi)\mathbf{E}_x \; = \; m\ddot{x}\mathbf{E}_x\,,$$

$$-R\mathbf{E}_y \times \left(N\mathbf{E}_y - \mu_d|N|\frac{\dot{x} + R\dot{\theta}}{|\dot{x} + R\dot{\theta}|}\mathbf{E}_x\right) \; = \; I_{zz}\ddot{\theta}\mathbf{E}_z\,.$$

From these equations we obtain differential equations for x and θ, and an equation for the normal force $N\mathbf{E}_y$.

First, for the normal force, we find that $N\mathbf{E}_y = mg\cos(\phi)\mathbf{E}_y$. Next, the differential equations are

$$-\mu_d mg\cos(\phi)\mathrm{sgn}(v_P) + mg\sin(\phi) \; = \; m\ddot{x}\,,$$
$$-\mu_d mgR\cos(\phi)\mathrm{sgn}(v_P) \; = \; I_{zz}\ddot{\theta}\,,$$

where

$$\mathrm{sgn}(v_P) = \frac{\dot{x} + R\dot{\theta}}{|\dot{x} + R\dot{\theta}|}\,,$$

and $v_P = \dot{x} + R\dot{\theta}$ is the slip velocity.

Solving these equations subject to the initial conditions $x(t = 0) = 0$, $\theta(t = 0) = 0$, $\dot{x}(t = 0) = v_0 > 0$, and $\dot{\theta}(t = 0) = \omega_0$, we find that

$$x(t) \; = \; \frac{g\cos(\phi)}{2}\left(\tan(\phi) - \mu_d\mathrm{sgn}(v_P)\right)t^2 + v_0 t\,,$$

$$\theta(t) \; = \; -\mu_d\left(\frac{mgR\cos(\phi)}{2I_{zz}}\right)\mathrm{sgn}(v_P)t^2 + \omega_0 t\,.$$

Initially, the body will move down the incline. At a later time, however, it is possible that $\mathbf{v}_P = v_P\mathbf{E}_x = \mathbf{0}$. In this case, the body will roll, and once it starts to roll, it will remain rolling.[12]

9.7   An Imbalanced Rotor

Many applications can be modeled as a rigid body rotating about a fixed axis. In particular, driveshafts in automobiles and turbomachinery. If the

---

[12]This type of transition occurs in bowling balls as they approach the pins. To aid this phenomenon, the bowling lanes are usually waxed to increase the friction between the ball and the lane as the former nears the pins. Some readers may also have noticed the same phenomenon in pool, a problem that is discussed in Article 239 of Routh [52].

rigid body is balanced, then the axis of rotation corresponds to a principal axis of the rigid body. However, this is very difficult, and expensive, to achieve. The problem we discuss below is classical. It illustrates the observed phenomenon that because a rigid body is not rotating about a principal axis, the bearing forces supporting it will be periodic, as opposed to constant if it were. The periodic nature of the bearing forces causes the bearings to fail in fatigue. Such a failure can be catastrophic.

The prototypical example of an imbalanced rigid body is shown in Figure 9.9. It consists of a homogeneous disk of mass m, radius R and thickness t which is welded to a homogeneous shaft of mass M, length $2L$, and radius r. The centers of mass of these rigid bodies are coincident, and the disk is inclined at an angle γ to the vertical. The rigid body, which consists of the shaft and the disk, is supported by bearings at A and B. This rigid body is often called a rotor. Finally, an applied torque $S\mathbf{E}_z$ acts on the shaft.

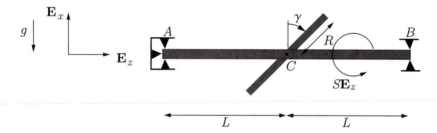

FIGURE 9.9. The imbalanced rotor problem

9.7.1   Kinematics

We assume that the motion is such that the axis of rotation is fixed: $\boldsymbol{\omega} = \dot{\theta}\mathbf{E}_z = \omega\mathbf{E}_z$ and $\boldsymbol{\alpha} = \ddot{\theta}\mathbf{E}_z$. The corotational basis is defined in the usual manner:

$$\mathbf{e}_x = \cos(\theta)\mathbf{E}_x + \sin(\theta)\mathbf{E}_y, \quad \mathbf{e}_y = -\sin(\theta)\mathbf{E}_x + \cos(\theta)\mathbf{E}_y, \quad \mathbf{e}_z = \mathbf{E}_z.$$

We also recall the relations

$$\dot{\mathbf{e}}_x = \dot{\theta}\mathbf{e}_y, \quad \dot{\mathbf{e}}_y = -\dot{\theta}\mathbf{e}_x, \quad \dot{\mathbf{e}}_z = \mathbf{0}.$$

We denote the position vector of the center of mass of the rigid body by

$$\bar{\mathbf{x}} = x_0\mathbf{E}_x + y_0\mathbf{E}_y + z_0\mathbf{E}_z.$$

Here, the bearings at A and B are such that $\bar{\mathbf{x}}$ is a constant:

$$\bar{\mathbf{v}} = \mathbf{0}, \quad \bar{\mathbf{a}} = \mathbf{0}.$$

Next, we address the angular momentum of the rigid body. Here, it is important to note that $\{\mathbf{e}_x, \mathbf{e}_y, \mathbf{e}_z\}$ are not principal axes. For a rigid body rotating about the $\mathbf{e}_z$ axis, we recall from Section 1.3 that

$$
\begin{aligned}
\mathbf{H} &= I_{xz}\omega\mathbf{e}_x + I_{yz}\omega\mathbf{e}_y + I_{zz}\omega\mathbf{E}_z\,, \\
\dot{\mathbf{H}} &= \left(I_{xz}\dot{\omega} - I_{yz}\omega^2\right)\mathbf{e}_x + \left(I_{yz}\dot{\omega} + I_{xz}\omega^2\right)\mathbf{e}_y + I_{zz}\dot{\omega}\mathbf{E}_z\,.
\end{aligned}
$$

For the rigid body of interest, a long, but straightforward, calculation shows that the inertias of interest are

$$
\begin{aligned}
I_{xz} &= \left(\frac{mt^2}{12} - \frac{mR^2}{4}\right)\sin(\gamma)\cos(\gamma)\,, \\
I_{yz} &= 0\,, \\
I_{zz} &= \frac{Mr^2}{2} + \left(\frac{mR^2}{4} + \frac{mt^2}{12}\right)\sin^2(\gamma) + \frac{mR^2}{2}\cos^2(\gamma)\,.
\end{aligned}
$$

Notice that if $\gamma = 0$, 90, 180, or 270 degrees, then I_{xz} vanishes and $\mathbf{E}_z$ is a principal axis of the rigid body.

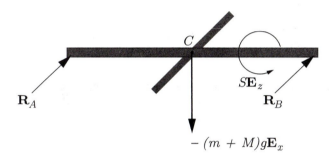

FIGURE 9.10. Free-body diagram

9.7.2 Forces and Moments

The free-body diagram is shown in Figure 9.10. The reaction forces at the bearings have the representations

$$
\mathbf{R}_A = R_{Ax}\mathbf{E}_x + R_{Ay}\mathbf{E}_y + R_{Az}\mathbf{E}_z\,, \quad \mathbf{R}_B = R_{Bx}\mathbf{E}_x + R_{By}\mathbf{E}_y\,.
$$

These 5 forces ensure that the center of mass remains fixed and that the angular velocity vector has components in the $\mathbf{E}_z$ direction only. There is also a gravitational force on the rigid body, $-(M + m)g\mathbf{E}_x$, in addition to an applied torque $S\mathbf{E}_z$. Consequently, the resultant force and moment acting on the rigid body are

$$
\mathbf{F} = \mathbf{R}_A + \mathbf{R}_B - (M + m)g\mathbf{E}_x\,, \quad \mathbf{M} = L\mathbf{E}_z \times (\mathbf{R}_B - \mathbf{R}_A) + S\mathbf{E}_z\,.
$$

9.7.3   Balance Laws

From the balance laws, we find that

$$\mathbf{R}_A + \mathbf{R}_B - (M+m)g\mathbf{E}_x = \mathbf{0},$$
$$L\mathbf{E}_z \times (\mathbf{R}_B - \mathbf{R}_A) + S\mathbf{E}_z = I_{xz}\dot{\omega}\mathbf{e}_x + I_{xz}\omega^2\mathbf{e}_y + I_{zz}\dot{\omega}\mathbf{E}_z.$$

This gives 6 equations for the 5 unknown components of the reaction forces and a differential equation for θ.

9.7.4   Analysis

We turn our attention to the case where $\boldsymbol{\omega}$ is a constant: $\boldsymbol{\omega} = \omega_0\mathbf{E}_z$. The 6 equations above simplify to

$$R_{Ax} + R_{Bx} - (M+m)g = 0,$$
$$R_{Ay} + R_{By} = 0,$$
$$R_{Az} = 0,$$
$$L(R_{Ay} - R_{By}) = -I_{xz}\omega_0^2\sin(\omega_0 t),$$
$$L(R_{Ax} - R_{Bx}) = I_{xz}\omega_0^2\cos(\omega_0 t),$$
$$S = 0.$$

Notice that from the $\mathbf{E}_z$ component of the balance of angular momentum we find that $S = 0$. That is, no applied torque is needed to rotate the rigid body at a constant angular speed. Solving for the bearing forces, we obtain

$$\mathbf{R}_A = \left(\frac{M+m}{2}\right)g\mathbf{E}_x - \left(\frac{I_{xz}\omega_0^2}{2L}\right)(\cos(\omega_0 t)\mathbf{E}_x + \sin(\omega_0 t)\mathbf{E}_y),$$
$$\mathbf{R}_B = \left(\frac{M+m}{2}\right)g\mathbf{E}_x + \left(\frac{I_{xz}\omega_0^2}{2L}\right)(\cos(\omega_0 t)\mathbf{E}_x + \sin(\omega_0 t)\mathbf{E}_y).$$

Clearly, these forces are the superposition of constant and periodic terms. Recalling the expression for I_{xz} given previously, then it is easy to see that the periodic component of these forces vanishes when $\gamma = 0, 90, 180$ or 270 degrees and the rigid body is then said to be balanced.

9.8   Summary

The first set of important results in this chapter pertained to a system of forces and a pure moment acting on a rigid body. Specifically, for a system of K forces $\mathbf{F}_i$ ($i = 1, \ldots, K$) and a moment $\mathbf{M}_p$, which is not due to the moment of an applied force, acting on the rigid body, the resultant force $\mathbf{F}$

and moments are

$$\mathbf{F} = \sum_{i=1}^{K} \mathbf{F}_i \,,$$

$$\mathbf{M}_O = \mathbf{M}_p + \sum_{i=1}^{K} \mathbf{x}_i \times \mathbf{F}_i \,,$$

$$\mathbf{M} = \mathbf{M}_p + \sum_{i=1}^{K} (\mathbf{x}_i - \bar{\mathbf{x}}) \times \mathbf{F}_i \,.$$

Here, $\mathbf{M}_O$ is the resultant moment relative to a fixed point O, while $\mathbf{M}$ is the resultant moment relative to the center of mass C of the rigid body.

The relationship between forces and moments and the motion of the rigid body is postulated using the balance laws. There are two equivalent sets of balance laws:

$$\mathbf{F} = m\dot{\bar{\mathbf{v}}} \,, \quad \mathbf{M}_O = \dot{\mathbf{H}}_O \,,$$

and

$$\mathbf{F} = m\dot{\bar{\mathbf{v}}} \,, \quad \mathbf{M} = \dot{\mathbf{H}} \,.$$

When these balance laws are specialized to the case of a fixed-axis rotation, the expressions for $\dot{\mathbf{H}}_O$ and $\dot{\mathbf{H}}$ simplify. For instance,

$$\mathbf{F} = m\dot{\bar{\mathbf{v}}} \,,$$
$$\mathbf{M} = \left(I_{xz}\dot{\omega} - I_{yz}\omega^2 \right) \mathbf{e}_x + \left(I_{yz}\dot{\omega} + I_{xz}\omega^2 \right) \mathbf{e}_y + I_{zz}\dot{\omega}\mathbf{E}_z \,.$$

In most problems, $\{\mathbf{E}_x, \mathbf{E}_y, \mathbf{E}_z\}$ are chosen such that $I_{xz} = I_{yz} = 0$.

To establish conservations of energy, two equivalent forms of the work-energy theorem were developed in Section 2. First, however, the Koenig decomposition for the kinetic energy of a rigid body was established:

$$T = \frac{1}{2}m\bar{\mathbf{v}} \cdot \bar{\mathbf{v}} + \frac{1}{2}\mathbf{H} \cdot \boldsymbol{\omega} \,.$$

This was then followed by a development of the work-energy theorem for a rigid body:

$$\frac{dT}{dt} = \mathbf{F} \cdot \bar{\mathbf{v}} + \mathbf{M} \cdot \boldsymbol{\omega} = \sum_{i=1}^{K} \mathbf{F}_i \cdot \mathbf{v}_i + \mathbf{M}_p \cdot \boldsymbol{\omega} \,.$$

To establish energy conservation results, this theorem is used in a similar manner to the one employed with particles and systems of particles.

Four important sets of applications were then discussed:

1. Purely translational motion of a rigid body where $\boldsymbol{\omega} = \boldsymbol{\alpha} = \mathbf{0}$.

2. A rigid body with a fixed point O.

3. Rolling rigid bodies and sliding rigid bodies.

4. Imbalanced rotors.

It is important to note that for the second set of applications, the balance law $\mathbf{M}_O = \dot{\mathbf{H}}_O$ is more convenient to use than $\mathbf{M} = \dot{\mathbf{H}}$. The role of $\mathbf{M}_c$ in these problems is to ensure that the axis of rotation remains $\mathbf{E}_z$. Finally, the four steps discussed in Section 1.4 are used as a guide to solving all of the applications.

9.9 Exercises

The following short exercises are intended to assist you in reviewing Chapter 9.

9.1 Starting from the definitions of $\mathbf{M}$ and $\mathbf{M}_O$, show that $\mathbf{M}_O = \dot{\mathbf{H}}_O$ along with $\mathbf{F} = m\dot{\mathbf{v}}$ implies that $\mathbf{M} = \dot{\mathbf{H}}$.

9.2 For the overturning cart discussed in Section 3.1, show that if the cart is tall (i.e., $a \gg b$) then the toppling force P is smaller than if the cart were stout (i.e., $a \ll b$).

9.3 Consider a cart with the same dimensions as the one discussed in Section 3.1. Suppose that the applied force $\mathbf{P} = \mathbf{0}$, but the front wheels are driven. The driving force on the respective front wheels is assumed to be

$$\mathbf{F}_1 = \mu N_{2y}\mathbf{E}_x\,, \quad \mathbf{F}_3 = \mu N_{3y}\mathbf{E}_x\,,$$

where μ is a constant. Calculate the resulting acceleration vector of the center of mass of the cart.

9.4 Consider the example of a rigid body rotating about a fixed point O discussed in Section 4. Starting from $\mathbf{H}_O = \mathbf{H} + \bar{\mathbf{x}} \times m\bar{\mathbf{v}}$, show that

$$\mathbf{H}_0 \cdot \boldsymbol{\omega} = \mathbf{H} \cdot \boldsymbol{\omega} + m\bar{\mathbf{v}} \cdot \bar{\mathbf{v}}\,.$$

Why is this result useful?

9.5 When solving the example discussed in Section 4, one person uses the balance law $\mathbf{M} = \dot{\mathbf{H}}$ instead of $\mathbf{M}_O = \dot{\mathbf{H}}_O$. Why is there nothing wrong with their approach?

9.6 As a special case of the example discussed in Section 4, consider a long slender rod of length L and mass m which is attached at one of its ends to a pin-joint. Show that $\mathbf{H} = \frac{1}{12}mL^2\dot{\theta}\mathbf{E}_z$, $\bar{\mathbf{v}} = \frac{L}{2}\dot{\theta}\mathbf{e}_y$, and $\mathbf{H}_O = \frac{1}{3}mL^2\dot{\theta}\mathbf{E}_z$.

9.7 For the problem discussed in Exercise 9.6, show that if the sole applied force acting on the rod is a gravitational force, $-mg\mathbf{E}_y$, the equation governing the motion of the rod is

$$\frac{mL^2}{3}\ddot{\theta} = -\frac{mgL}{2}\cos(\theta).$$

Furthermore, prove that the total energy of the rod is conserved. Why is this problem analogous to a planar pendulum problem?

9.8 Consider a rigid body which is rolling on a fixed surface. Suppose that, apart from the friction and normal forces at the point of contact P, the applied forces acting on the body are conservative, then why is the total energy of the rigid body conserved?

9.9 Using the static friction criterion discussed in Section 6.3, show that a circular disk of mass m and radius R can roll without slipping on a steeper incline than a circular hoop of the same mass and radius.

9.10 For a rolling disk on a horizontal incline, show that $|\dot{x}| \leq \mu_s mg$. Similarly, for a sliding disk show that $|\dot{x}| = \mu_d mg$. Using, these results, explain why rolling disks decelerate faster than sliding disks. This observation is the reason for the desirability anti-lock-braking-systems (ABS) in automobiles.

9.11 Consider the sliding rigid disk discussed in Section 6.4 and suppose that $\phi = 0$. Determine $\theta(t)$ and $x(t)$ for the two cases where initially $\mathrm{sgn}(v_P) > 0$ and $\mathrm{sgn}(v_P) < 0$.

9.12 Recall the imbalanced rotor discussed in Section 7. For various values of ω_0, plot the components R_{Ax}, R_{Bx}, R_{Ay}, and R_{By} using the numerical values $L = 10$ meters, $m = 20$ kilograms, $M = 100$ kilograms, $\gamma = -0.01$ radians, $t = 0.01$ meters, and $R = 1$ meter.

10

Systems of Particles and Rigid Bodies

TOPICS

This chapter is the culmination of the primer. To start, the linear momentum of a system of K particles and N rigid bodies is discussed. Similarly, the angular momenta and kinetic energy of such a system are developed. We then turn to the balance laws for such a system. The complete analysis of the resulting differential equations that these laws provide is usually beyond the scope of an undergraduate engineering dynamics course, and instead we focus on some particular results. These results involve using conservations of energy, angular momentum, and linear momentum. We then illustrate how one or more such conservations can be used to obtain solutions to some problems.

10.1 A System of Particles and Rigid Bodies

The problems of interest feature one or more rigid bodies, which may possibly be interacting with one or more particles. To cover as many cases as possible, we now discuss a particular system of K particles and N rigid bodies. It is general enough to cater to all of our subsequent developments and examples.

We use the index k to denote each particle ($k = 1, \ldots, K$). The mass of the kth particle is denoted by $_pm_n$, its position vector relative to a fixed origin O by $\mathbf{r}_k$, and the resultant external force acting on the particle is denoted by $_p\mathbf{F}_k$ (see Figure 10.1).

Particle of mass $_p m_k$

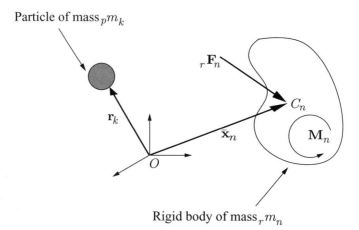

Rigid body of mass $_r m_n$

FIGURE 10.1. The kth particle and the nth rigid body

Similarly, we use the index n for each rigid body $(n = 1, \ldots, N)$. The mass of the nth rigid body is denoted by $_r m_n$, the position vector of its center of mass C_n relative to a fixed origin O by $\bar{\mathbf{x}}_n$, its angular velocity vector by $\boldsymbol{\omega}_n$, and the resultant external force and moment (relative to its center of mass) acting on the rigid body are denoted by $_r\mathbf{F}_n$ and $_r\mathbf{M}_n$, respectively (see Figure 10.1).

The center of mass C of this system has a position vector $\bar{\mathbf{x}}$, which is defined by

$$\bar{\mathbf{x}} = \frac{1}{m} \left(\sum_{k=1}^{K} {}_p m_k \mathbf{r}_k + \sum_{n=1}^{N} {}_r m_n \bar{\mathbf{x}}_n \right),$$

where m is the total mass of the system:

$$m = \sum_{k=1}^{K} {}_p m_k + \sum_{n=1}^{N} {}_r m_n.$$

You should notice how this definition is an obvious extension to others that you have seen previously.

10.1.1   Momenta and Kinetic Energy

The linear momentum $\mathbf{G}$ of the system is the sum of the linear momenta of the individual particles and rigid bodies:

$$\mathbf{G} = \sum_{k=1}^{K} {}_p m_k \dot{\mathbf{r}}_k + \sum_{n=1}^{N} {}_r m_n \dot{\bar{\mathbf{x}}}_n.$$

Notice that the linear momentum of the system can also be expressed in terms of the velocity vector of the center of mass of the system and the

total mass of the system:
$$\mathbf{G} = m\dot{\bar{\mathbf{x}}}.$$

This result follows from the definition of the center of mass.

The angular momentum $\mathbf{H}_O$ of the system relative to the fixed point O is the sum of the individual angular momenta relative to O:

$$\mathbf{H}_O = \sum_{k=1}^{K} \mathbf{r}_k \times {}_p m_k \dot{\mathbf{r}}_k + \sum_{n=1}^{N} \mathbf{H}_n + \bar{\mathbf{x}}_n \times {}_r m_n \dot{\bar{\mathbf{x}}}_n.$$

Similarly, the angular momentum $\mathbf{H}$ of the system relative to its center of mass is

$$\mathbf{H} = \sum_{k=1}^{K} (\mathbf{r}_k - \bar{\mathbf{x}}) \times {}_p m_k \dot{\mathbf{r}}_k + \sum_{n=1}^{N} \mathbf{H}_n + (\bar{\mathbf{x}}_n - \bar{\mathbf{x}}) \times {}_r m_n \dot{\bar{\mathbf{x}}}_n.$$

In both of the previous equations, $\mathbf{H}_n$ is the angular momentum of the nth rigid body relative to its center of mass. You should notice that we have used the identities

$$\mathbf{H}_{On} = \mathbf{H}_n + \bar{\mathbf{x}}_n \times {}_r m_n \dot{\bar{\mathbf{x}}}_n,$$

where $\mathbf{H}_{On}$ is the angular momentum of the nth rigid body relative to O.

Finally, the kinetic energy T of the system is defined to be the sum of the kinetic energies of its constituents:

$$T = \sum_{k=1}^{K} \frac{1}{2} {}_p m_k \dot{\mathbf{r}}_k \cdot \dot{\mathbf{r}}_k + \sum_{n=1}^{N} \frac{1}{2} {}_r m_n \dot{\bar{\mathbf{x}}}_n \cdot \dot{\bar{\mathbf{x}}}_n + \frac{1}{2} \mathbf{H}_n \cdot \boldsymbol{\omega}_n.$$

10.1.2 Impulses, Momenta, and Balance Laws

For each rigid body of the system, one has the balance of linear momentum and the balance of angular momentum:

$${}_r \mathbf{F}_n = {}_r m_n \ddot{\bar{\mathbf{x}}}_n, \quad \mathbf{M}_{On} = \dot{\mathbf{H}}_{On}, \quad (n = 1, \ldots, N).$$

Here, $\mathbf{M}_{On}$ is the resultant moment, relative to the point O, acting on the nth rigid body. We also have an alternative form of the balance of angular momentum relative to the center of mass of the nth rigid body: $\mathbf{M}_n = \dot{\mathbf{H}}_n$. In addition, for each particle one has the balance of linear momentum:

$${}_p \mathbf{F}_k = {}_p m_k \ddot{\mathbf{r}}_k, \quad (k = 1, \ldots, K).$$

As discussed in Chapter 6, the angular momentum theorem for a particle is derived from this balance law. Hence, for a particle, the balance of angular momentum is not a separate postulate as it is with rigid bodies.

Let $\mathbf{M}_O$ denote the resultant moment acting on the system relative to the point O, $\mathbf{M}$ denote the resultant moment acting on the system relative to its center of mass, and $\mathbf{F}$ denote the resultant force acting on the system. Then, adding the balances of linear momenta, we find that

$$\mathbf{F} = \dot{\mathbf{G}},$$

where

$$\mathbf{F} = \sum_{k=1}^{K} {}_p\mathbf{F}_k + \sum_{n=1}^{N} {}_r\mathbf{F}_n.$$

Similarly, adding the balances of angular momenta relative to the point O,

$$\mathbf{M}_O = \dot{\mathbf{H}}_O,$$

where

$$\mathbf{M}_O = \sum_{k=1}^{K} {}_p\mathbf{r}_k \times {}_p\mathbf{F}_k + \sum_{n=1}^{N} \mathbf{M}_{On}.$$

Hence, we have balances of linear and angular momenta for the system.[1]

Another form of the balance laws can be obtained by integrating both sides of $\mathbf{F} = \dot{\mathbf{G}}$, $\mathbf{M} = \dot{\mathbf{H}}$, and $\mathbf{M}_O = \dot{\mathbf{H}}_O$. These forms are known as the *impulse-momentum forms* or the *integral forms* of the balance laws:

$$\mathbf{G}(t) - \mathbf{G}(t_0) = \int_{t_0}^{t} \mathbf{F}(\tau)d\tau,$$

$$\mathbf{H}(t) - \mathbf{H}(t_0) = \int_{t_0}^{t} \mathbf{M}(\tau)d\tau,$$

$$\mathbf{H}_O(t) - \mathbf{H}_O(t_0) = \int_{t_0}^{t} \mathbf{M}_O(\tau)d\tau.$$

Here, the integral of the force $\mathbf{F}$ is known as the *linear impulse* (of $\mathbf{F}$), and the integral of the moment $\mathbf{M}_O$ is known as the *angular impulse* (of $\mathbf{M}_O$). As you have already seen, this form of the balance laws is extremely useful in analyzing impact problems, where the impulse of certain forces and moments dominates contributions from other forces and moments.

Although we did not discuss the impulse-momentum forms of the balance laws for rigid bodies, it should be obvious that for each rigid body, one has

$$_r m_n \bar{\mathbf{v}}_n(t) - {}_r m_n \bar{\mathbf{v}}_n(t_0) = \int_{t_0}^{t} {}_r\mathbf{F}_n(\tau)d\tau,$$

$$\mathbf{H}_{On}(t) - \mathbf{H}_{On}(t_0) = \int_{t_0}^{t} \mathbf{M}_{On}(\tau)d\tau, \quad (n = 1, \ldots, N).$$

---

[1]You should be able to consider special cases of these results, for example, cases where the system of interest contains either no rigid bodies or no particles. We could establish a balance of angular momentum relative to the center of mass of the system ($\mathbf{M} = \dot{\mathbf{H}}$), but we leave this as an exercise. Such an exercise will involve using $\mathbf{F} = \dot{\mathbf{G}}$ and $\mathbf{M}_O = \dot{\mathbf{H}}_O$. Its proof is similar to the case of a single rigid body.

The corresponding results for a single particle were discussed in Chapter 6.

10.1.3 Conservations

We now address the possibility that in certain problems, a component of the linear momentum $\mathbf{G}$ of the system, a component of the angular momenta $\mathbf{H}_O$ or $\mathbf{H}$ of the system, and/or the total energy E of the system is conserved. We have developed these results three times previously: once for a single particle, once for a system of particles, and once for a single rigid body.

Conservation of Linear Momentum

Let us first deal with conservation of linear momentum. Given a vector $\mathbf{c} = \mathbf{c}(t)$, then it is easily seen that for $\mathbf{G} \cdot \mathbf{c}$ to be constant during the motion of the system, we must have $\mathbf{F} \cdot \mathbf{c} + \mathbf{G} \cdot \dot{\mathbf{c}} = 0$. Finding the vector $\mathbf{c}$ that satisfies $\mathbf{F} \cdot \mathbf{c} + \mathbf{G} \cdot \dot{\mathbf{c}} = 0$ for a particular system is an art, and we shall discuss examples shortly. You should notice that if $\mathbf{F} = \mathbf{0}$, then $\mathbf{G}$ is conserved.

The conservation of components of $\mathbf{G}$ is often used in impact problems.

Conservation of Angular Momentum

Next, we have conservation of angular momentum. Given a vector $\mathbf{c} = \mathbf{c}(t)$, we seek to determine when $\mathbf{H}_O \cdot \mathbf{c}$ is constant during the motion of the system. To do this, we calculate

$$\frac{d}{dt} (\mathbf{H}_O \cdot \mathbf{c}) = \dot{\mathbf{H}}_O \cdot \mathbf{c} + \mathbf{H}_O \cdot \dot{\mathbf{c}} = \mathbf{M}_O \cdot \mathbf{c} + \mathbf{H}_O \cdot \dot{\mathbf{c}}.$$

It follows that it is necessary and sufficient for $\mathbf{M}_O \cdot \mathbf{c} + \mathbf{H}_O \cdot \dot{\mathbf{c}} = 0$ for $\mathbf{H}_O \cdot \mathbf{c}$ to be conserved during the motion of the system. Again, finding $\mathbf{c}$ such that $\mathbf{M}_O \cdot \mathbf{c} + \mathbf{H}_O \cdot \dot{\mathbf{c}} = 0$ for a given system is an art. You should notice that if $\mathbf{M}_O = \mathbf{0}$, then $\mathbf{H}_O$ is conserved. The corresponding results for $\mathbf{H}$ are easily inferred.

The most common occurrence of conservation of angular momentum is a system of interconnected rigid bodies in space. There, because one assumes that $\mathbf{M} = \mathbf{0}$, the angular momentum $\mathbf{H}$ is conserved, and astronauts use this conservation to change their orientation during space walks. Specifically, one can consider an astronaut as a system of rigid bodies. By changing the relative orientation of these bodies, they change the angular velocity vectors of parts of their bodies, and in this manner change their overall orientation. A similar principle is behind the falling cat, which "always" seems to land on its feet.[2]

Conservation of Energy

Finally, we turn to the conservation of energy. We start with the definition of the kinetic energy of the system, and, by using the work-energy theorems for the individual particles and rigid bodies, we establish a work-energy theorem for the system. This work-energy theorem was used, as in particles, systems of particles, and single rigid bodies, to establish whether or not the total energy of the system is conserved. We now proceed to establish the work-energy theorem for the system.

Recall that

$$T = \sum_{k=1}^{K} \frac{1}{2}{}_p m_k \dot{\mathbf{r}}_k \cdot \dot{\mathbf{r}}_k + \sum_{n=1}^{N} \frac{1}{2}{}_r m_n \dot{\bar{\mathbf{x}}}_n \cdot \dot{\bar{\mathbf{x}}}_n + \frac{1}{2}\mathbf{H}_n \cdot \boldsymbol{\omega}_n .$$

Taking the derivative of this expression and using the work-energy theorems discussed previously,

$$\frac{d}{dt}\left(\frac{1}{2}{}_p m_k \dot{\mathbf{r}}_k \cdot \dot{\mathbf{r}}_k\right) = {}_p\mathbf{F}_k \cdot \dot{\mathbf{r}}_k , \quad (k=1,\dots,K) ,$$

$$\frac{d}{dt}\left(\frac{1}{2}{}_r m_n \dot{\bar{\mathbf{x}}}_n \cdot \dot{\bar{\mathbf{x}}}_n + \frac{1}{2}\mathbf{H}_n \cdot \boldsymbol{\omega}_n\right) = {}_r\mathbf{F}_n \cdot \dot{\bar{\mathbf{x}}}_n$$
$$+ \mathbf{M}_n \cdot \boldsymbol{\omega}_n , \quad (n=1,\dots,N) ,$$

we find that

$$\frac{dT}{dt} = \sum_{k=1}^{K} {}_p\mathbf{F}_k \cdot \dot{\mathbf{r}}_k + \sum_{n=1}^{N} {}_r\mathbf{F}_n \cdot \dot{\bar{\mathbf{x}}}_n + \mathbf{M}_n \cdot \boldsymbol{\omega}_n .$$

This is the work-energy theorem for the system. Starting from this theorem and substituting for the forces and moments on the individual constituents of the system, one can ascertain whether or not the total energy of the system is conserved. Again, this procedure is identical to the one previously discussed. You should also note that for each rigid body we have an alternative form of the work-energy theorem that we can use in place of the terms on the right-hand side of the above equation.[3] We will do this in the example below.

---

[2]See the photos for the falling cat in Crabtree [17] and Kane and Scher [34]. References to modern approaches to this problem can be found in Shapere and Wilczek [57] and Fecko [25].

[3]See Section 2.3 of Chapter 9.

10.2   An Example of Two Rigid Bodies

Here, as our first example, we consider two connected rigid bodies (cf. Figure 10.2). One of the bodies is a slender rod of length L and mass m_1 that is pin-jointed at O. This point is fixed. The other is a circular disk of mass m_2 and radius R. They are connected by a pin-joint at the point A that lies at the outer extremity of the rod and the edge of the disk. Both bodies rotate about the $\mathbf{E}_z$ axis, and this axis is also a principal axis for both bodies. A gravitational force acts on the bodies.

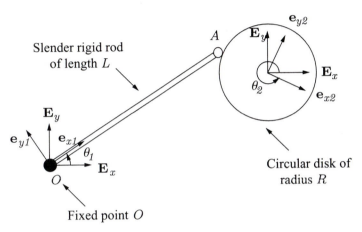

FIGURE 10.2. Two coupled rigid bodies

We wish to determine which kinematical quantities are conserved for this system. At the end of this section we will discuss related problems and questions concerning this system.

10.2.1   Kinematics

To proceed, we first calculate the linear momentum $\mathbf{G}$, angular momentum $\mathbf{H}_O$, and kinetic energy T of the system. It is convenient to define two corotational bases $\{\mathbf{e}_{x1}, \mathbf{e}_{y1}, \mathbf{e}_{z1} = \mathbf{E}_z\}$ and $\{\mathbf{e}_{x2}, \mathbf{e}_{y2}, \mathbf{e}_{z2} = \mathbf{E}_z\}$. You should notice that these bases are related by

$$\begin{aligned}
\mathbf{e}_{x2} &= \cos(\theta_2 - \theta_1)\,\mathbf{e}_{x1} + \sin(\theta_2 - \theta_1)\,\mathbf{e}_{y1}\,, \\
\mathbf{e}_{y2} &= \cos(\theta_2 - \theta_1)\,\mathbf{e}_{y1} - \sin(\theta_2 - \theta_1)\,\mathbf{e}_{x1}\,.
\end{aligned}$$

Since $\mathbf{E}_z$ is a principal axis, we can use our results from Section 8 of Chapter 8 to write

$$\mathbf{H}_1 = \frac{m_1 L^2}{12}\dot{\theta}_1\mathbf{E}_z\,, \quad \mathbf{H}_2 = \frac{m_2 R^2}{2}\dot{\theta}_2\mathbf{E}_z\,.$$

You should notice that the angular velocity vectors of the bodies are

$$\boldsymbol{\omega}_1 = \dot{\theta}_1\mathbf{E}_z\,, \quad \boldsymbol{\omega}_2 = \dot{\theta}_2\mathbf{E}_z\,.$$

These velocities were used to calculate the angular momenta of the bodies relative to their centers of mass.

One also has the following representations:[4]

$$\bar{\mathbf{x}}_1 = \frac{L}{2}\mathbf{e}_{x1}, \quad \bar{\mathbf{x}}_2 = L\mathbf{e}_{x1} + R\mathbf{e}_{x2}, \quad \mathbf{x}_A = L\mathbf{e}_{x1}.$$

Differentiating these representations, we find that

$$\bar{\mathbf{v}}_1 = \frac{L}{2}\dot{\theta}_1\mathbf{e}_{y1}, \quad \bar{\mathbf{v}}_2 = L\dot{\theta}_1\mathbf{e}_{y1} + R\dot{\theta}_2\mathbf{e}_{y2}, \quad \mathbf{v}_A = L\dot{\theta}_1\mathbf{e}_{y1}.$$

Hence, the linear momentum of the system is

$$\mathbf{G} = (m_1 + 2m_2)\frac{L}{2}\dot{\theta}_1\mathbf{e}_{y1} + m_2 R\dot{\theta}_2\mathbf{e}_{y2}.$$

The angular momentum of the system relative to O is, by definition,

$$\mathbf{H}_O = \mathbf{H}_1 + \bar{\mathbf{x}}_1 \times m_1\bar{\mathbf{v}}_1 + \mathbf{H}_2 + \bar{\mathbf{x}}_2 \times m_2\bar{\mathbf{v}}_2.$$

Substituting for the kinematical quantities on the right-hand side of this equation, one obtains, after a substantial amount of algebra,

$$\mathbf{H}_O = \frac{m_1 L^2}{3}\dot{\theta}_1\mathbf{E}_z + m_2\left(L^2\dot{\theta}_1 + \frac{3R^2}{2}\dot{\theta}_2\right)\mathbf{E}_z$$
$$+ m_2 RL\left(\dot{\theta}_2 + \dot{\theta}_1\right)\cos(\theta_2 - \theta_1)\mathbf{E}_z.$$

Finally, the kinetic energy of the system is

$$T = \frac{1}{2}m_1\bar{\mathbf{v}}_1 \cdot \bar{\mathbf{v}}_1 + \frac{1}{2}\mathbf{H}_1 \cdot \boldsymbol{\omega}_1 + \frac{1}{2}m_2\bar{\mathbf{v}}_2 \cdot \bar{\mathbf{v}}_2 + \frac{1}{2}\mathbf{H}_2 \cdot \boldsymbol{\omega}_2.$$

All of the ingredients are present to write this expression in terms of the kinematical quantities discussed earlier:

$$T = \frac{m_1 L^2}{6}\dot{\theta}_1^2 + \frac{m_2}{2}\left(L^2\dot{\theta}_1^2 + \frac{3R^2}{2}\dot{\theta}_2^2 + 2RL\dot{\theta}_1\dot{\theta}_2\cos(\theta_2 - \theta_1)\right).$$

10.2.2 Forces and Moments

In order to examine which conserved quantities are present in this system, we first need to determine the forces and moments acting on each body. These are summarized in the free-body diagrams shown in Figure 10.3. You should notice that there is a reaction force $\mathbf{R}_1 = R_{1x}\mathbf{E}_x + R_{1y}\mathbf{E}_y + R_{1z}\mathbf{E}_z$ at O, equal and opposite reaction forces of the form $\mathbf{R}_2 = R_{2x}\mathbf{E}_x + R_{2y}\mathbf{E}_y + R_{2z}\mathbf{E}_z$ at A, a reaction moment at O, $\mathbf{M}_{c1} = M_{c1x}\mathbf{E}_x + M_{c1y}\mathbf{E}_y$, and equal and opposite reaction moments of the form $\mathbf{M}_{c2} = M_{c2x}\mathbf{E}_x + M_{c2y}\mathbf{E}_y$ at

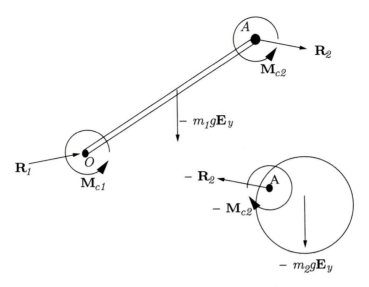

FIGURE 10.3. Free-body diagrams of the rod and the disk

A, These reactions ensure that the bodies are connected and their angular velocities are in the $\mathbf{E}_z$ direction.

In summary, the resultant forces are

$$
\begin{aligned}
\mathbf{F}_1 &= \mathbf{R}_1 + \mathbf{R}_2 - m_1 g \mathbf{E}_y , \\
\mathbf{F}_2 &= -\mathbf{R}_2 - m_2 g \mathbf{E}_y , \\
\mathbf{F} &= \mathbf{R}_1 - (m_1 + m_2) \, g \mathbf{E}_y .
\end{aligned}
$$

The resultant moments are

$$
\begin{aligned}
\mathbf{M}_{O1} &= \bar{\mathbf{x}}_1 \times (2\mathbf{R}_2 - m_1 g \mathbf{E}_y) + \mathbf{M}_{c1} + \mathbf{M}_{c2} , \\
\mathbf{M}_{O2} &= -2\bar{\mathbf{x}}_1 \times \mathbf{R}_2 - \bar{\mathbf{x}}_2 \times m_2 g \mathbf{E}_y - \mathbf{M}_{c2} , \\
\mathbf{M}_O &= \mathbf{M}_{O1} + \mathbf{M}_{O2} \\
&= (\bar{\mathbf{x}}_1 \times -m_1 g \mathbf{E}_y) + (\bar{\mathbf{x}}_2 \times -m_2 g \mathbf{E}_y) + \mathbf{M}_{c1} .
\end{aligned}
$$

10.2.3 Balance Laws and Analysis

Next, we examine conservation results. Clearly, $\mathbf{F} \neq \mathbf{0}$, even if gravity were absent, so $\mathbf{G}$ cannot be conserved. Next, we have that

$$
\dot{\mathbf{H}}_O = \mathbf{M}_O = (\bar{\mathbf{x}}_1 \times -m_1 g \mathbf{E}_y) + (\bar{\mathbf{x}}_2 \times -m_2 g \mathbf{E}_y) + \mathbf{M}_{c1} .
$$

Since $\mathbf{M}_O$ has components in the $\mathbf{E}_x$, $\mathbf{E}_y$, and $\mathbf{E}_z$ directions, no component of $\mathbf{H}_O$ is conserved. If gravity were absent however, then $\dot{\mathbf{H}}_O = \mathbf{M}_O = \mathbf{M}_{c1} = M_{c1x}\mathbf{E}_x + M_{c1y}\mathbf{E}_y$, and $\mathbf{H}_O \cdot \mathbf{E}_z$ would be conserved.

---

[4] For ease of notation, we drop the subscripts r and p used in the previous sections.

We now turn to the question of whether or not energy is conserved. To proceed, we start with the work-energy theorem for the system:

$$\frac{dT}{dt} = \mathbf{F}_1 \cdot \dot{\mathbf{x}}_1 + \mathbf{M}_1 \cdot \boldsymbol{\omega}_1 + \mathbf{F}_2 \cdot \dot{\mathbf{x}}_2 + \mathbf{M}_1 \cdot \boldsymbol{\omega}_2 .$$

Substituting for the forces and moments listed above, we find with some rearranging that[5]

$$\begin{aligned}
\frac{dT}{dt} &= \mathbf{R}_1 \cdot \mathbf{0} + \mathbf{R}_2 \cdot \mathbf{v}_A - m_1 g \mathbf{E}_y \cdot \bar{\mathbf{v}}_1 + (\mathbf{M}_{c1} + \mathbf{M}_{c2}) \cdot \boldsymbol{\omega}_1 \\
&\quad - \mathbf{R}_2 \cdot \mathbf{v}_A - m_2 g \mathbf{E}_y \cdot \bar{\mathbf{v}}_2 + -\mathbf{M}_{c2} \cdot \boldsymbol{\omega}_2 \\
&= -m_1 g \mathbf{E}_y \cdot \bar{\mathbf{v}}_1 - m_2 g \mathbf{E}_y \cdot \bar{\mathbf{v}}_2 .
\end{aligned}$$

Notice that we used the fact that the reaction moments are normal to the angular velocities, and, consequently, they do not contribute to the rate of change of kinetic energy. Manipulating the gravitational terms as usual, and performing some obvious cancellations, we find that the rate of change of the total energy,

$$E = T + m_1 g \mathbf{E}_y \cdot \bar{\mathbf{x}}_1 + m_2 g \mathbf{E}_y \cdot \bar{\mathbf{x}}_2 ,$$

is

$$\frac{dE}{dt} = 0 .$$

In summary, if gravity is present, then only the total energy is conserved. On the other hand, if gravity were absent, then, in addition to energy conservation, $\mathbf{H}_O \cdot \mathbf{E}_z$ would also be conserved.

10.2.4 A Related Example

A related problem is to assume that the system is in motion. At some time $t = t_1$ the pin-joint at A freezes up, so that $\dot{\theta}_1 = \dot{\theta}_2$ for $t > t_1$. Given $\theta_1(t_{1-})$, $\theta_2(t_{1-})$, $\dot{\theta}_1(t_{1-})$ and $\dot{\theta}_2(t_{1-})$, where t_{1-} denotes the instant just before the freeze-up, is it possible to determine $\dot{\theta}_1(t_{1+}) = \dot{\theta}_2(t_{1+})$, where t_{1+} denotes the instant just after the freeze-up?

The answer is yes! During the freeze-up, one can ignore the angular impulse due to gravity.[6] Then, from the integral form of the balance of angular momentum for the system, one has

$$\mathbf{H}_O(t_{1-}) \cdot \mathbf{E}_z = \mathbf{H}_O(t_{1+}) \cdot \mathbf{E}_z ,$$

---

[5] Notice that after the rearranging, one has the alternative form of the work-energy theorem. For a single rigid body, this alternative form was discussed in Section 2.3 of Chapter 9.

[6] Essentially, one is assuming that the freeze-up occurs instantaneously.

and this enables one to determine $\omega = \dot{\theta}_1(t_{1+}) = \dot{\theta}_2(t_{1+})$. Explicitly,

$$
\begin{aligned}
\mathbf{H}_O(t_{1-}) \cdot \mathbf{E}_z \;=\; & \frac{m_1 L^2}{3}\dot{\theta}_1(t_{1-}) \\
& + m_2\left(L^2 \dot{\theta}_1(t_{1-}) + \frac{3R^2}{2}\dot{\theta}_2(t_{1-}) \right) \\
& + m_2 RL\left(\dot{\theta}_2(t_{1-}) + \dot{\theta}_1(t_{1-}) \right) \cos\left(\theta_2(t_{1-}) - \theta_1(t_{1-}) \right), \\
\mathbf{H}_O(t_{1+}) \cdot \mathbf{E}_z \;=\; & \frac{m_1 L^2}{3}\omega + m_2\left(L^2 + \frac{3R^2}{2} \right)\omega \\
& + 2m_2 RL \cos\left(\theta_2(t_{1-}) - \theta_1(t_{1-}) \right)\omega.
\end{aligned}
$$

Equating these two expressions provides an equation to determine ω. One can also easily show that the energy is not conserved in this freeze-up, but we leave this as an exercise.

10.3   Impact of a Particle and a Rigid Body

We outline here some examples involving particles colliding with rigid bodies. As in collisions of particles with each other, one must be given some additional information: the coefficient of restitution for the problem at hand. Often, this is given implicitly. If the particle sticks to the body after the collision, one has that $e = 0$. If one is given the coefficient of restitution for these problems, it is important to note that one of the velocity vectors pertains to the particle, while the other pertains to the velocity vector of the material point of the body that the particle impacts.

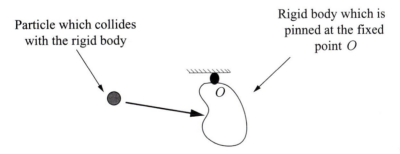

Particle which collides
with the rigid body

Rigid body which is
pinned at the fixed
point O

FIGURE 10.4. The first type of generic impact problem

Our discussion here is brief, and we provide few analytical details. Indeed, what makes these problems time-consuming is the calculation of linear and angular momenta before and after the impact. There are two types of problems common to undergraduate engineering courses in dynamics (see

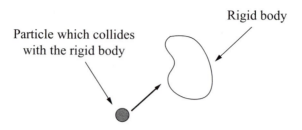

Rigid body

Particle which collides
with the rigid body

FIGURE 10.5. The second type of generic impact problem

Figures 10.4 and 10.5). Normally, energy is not conserved in these problems.

The first type of problem is where the rigid body is attached, usually, by a pin-joint to a fixed point O (see Figure 10.4). For these problems, one assumes that the collision is instantaneous, and hence the angular momentum component $\mathbf{H}_O \cdot \mathbf{E}_z$ is conserved, where $\mathbf{H}_O$ is the angular momentum of the system relative to O. That is, one ignores any gravitational impulses during the collision. The linear momentum $\mathbf{G}$ of the system is not conserved, because one has an impulse due to the reaction force at O. One then has a single equation: $\mathbf{H}_O \cdot \mathbf{E}_z$ is conserved. When the impact is such that the particle coalesces with the rigid body, this equation is usually used to determine the angular velocity of the particle-rigid body system immediately after the impact.

In the second type of problem, the rigid body is free to move on a plane, or in space (see Figure 10.5). Again, one assumes that the collision is instantaneous, and hence the angular momentum component $\mathbf{H} \cdot \mathbf{E}_z$ is conserved. Here, $\mathbf{H}$ is the angular momentum of the system relative to the center of mass of the system. The linear momentum $\mathbf{G}$ of the system is also conserved for these problems. Hence, one has 4 conservations. When the impact is such that the particle coalesces with the rigid body, these 4 conservations are normally used to determine the velocity vector of the center of mass and angular velocity vector of the particle-rigid body system immediately after the impact.

10.4   Beyond this Primer

The scope of the present book was influenced by what is expected of a student in an undergraduate engineering dynamics course. What is beyond its scope are areas of active research. We now turn to two of these areas.

First, many readers will notice the emphasis of establishing differential equations for the motion of the system. The nature of the solutions to these equations contain many of the predictions of the model developed for the system. These solutions are therefore a crucial component of verifying any model. Not surprisingly, there has been an enormous amount of research on the solutions to differential equations which arise from the development of models for mechanical systems. Fueled by recent advances in numerical computations, this is still an active research area (see, for example, Strogatz [62]). Based on our own experience, the predictions made by models of mechanical systems have usually been a source of enlightenment and improved understanding.

Another active research area is the development of more realistic models for impacting rigid bodies. These models are the basis for numerous simulations of vehicle collisions, and are also important in other applications. Of particular interest are models which incorporate frictional forces. The difficulties of establishing such theories was made evident in 1895 by Painlevé's paradoxical example of a rod sliding on a rigid horizontal surface [46].[7] Ruina [54] has also presented some interesting examples which illustrate some difficulties associated with the Coulomb friction laws we discussed in Chapter 4. We refer the interested reader to the review article by Stewart [60] for recent developments in this area.

10.5 Summary

This chapter has evident similarities to Chapter 7 where corresponding results for a system of particles were discussed. Rather than summarizing the main results presented in this Chapter 10, it is probably more useful to give a verbal outline of how they were established.

First, for a system of particles and rigid bodies, the center of mass is calculated using the masses of the constituents, the position vectors of the particles, and the position vectors of the centers of mass of the rigid bodies. The linear momentum $\mathbf{G}$ is calculated by summing the linear momenta of each of the constituents. This momentum is equal to the linear momentum of the center of mass of the system. Similarly, the angular momentum $\mathbf{H}_O$ is calculated by summing the angular momenta relative to O of the constituents. A similar remark applies for $\mathbf{H}$. Finally, the kinetic energy T of the system is calculated by summing the kinetic energies of the constituents. In general, $\mathbf{H}_O$ is not simply equal the angular momentum of the center of mass of the system, and T is not simply equal to the kinetic energy of the center of mass of the system.

As in systems of particles, one can add the balance laws for the system of particles and rigid body to arrive at the following balance laws:

$$\mathbf{F} = m\dot{\mathbf{v}}, \quad \mathbf{M} = \dot{\mathbf{H}}, \quad \mathbf{M}_O = \dot{\mathbf{H}}_O.$$

---

[7]Paul Painlevé (1863–1933) was a French mathematician and politician. It is interesting to note that he is credited as being the first airplane passenger of Wilbur Wright in 1908.

These laws are useful when establishing conservations of linear and angular momenta for the system of particles and rigid bodies. In addition, one can formulate the following integral forms of the balance laws:

$$\mathbf{G}(t) - \mathbf{G}(t_0) = \int_{t_0}^{t} \mathbf{F}(\tau)d\tau,$$

$$\mathbf{H}(t) - \mathbf{H}(t_0) = \int_{t_0}^{t} \mathbf{M}(\tau)d\tau,$$

$$\mathbf{H}_O(t) - \mathbf{H}_O(t_0) = \int_{t_0}^{t} \mathbf{M}_O(\tau)d\tau.$$

These results are very useful in impact problems.

Finally, the work-energy theorem for the system of particles and rigid bodies is obtained by adding the corresponding theorems for each of the constituents:

$$\frac{dT}{dt} = \sum_{k=1}^{K} {}_p\mathbf{F}_k \cdot \dot{\mathbf{r}}_k + \sum_{n=1}^{N} {}_r\mathbf{F}_n \cdot \dot{\bar{\mathbf{x}}}_n + \mathbf{M}_n \cdot \boldsymbol{\omega}_n.$$

As always, this theorem is useful for establishing conservation of energy results.

The main examples discussed in this chapter were a system of two rigid bodies and several impact problems. It is crucial to remember that the correct solution of these problems depends on one's ability to establish expressions for $\mathbf{G}$, $\mathbf{H}_O$, and $\mathbf{H}$.

10.6 Exercises

The following short exercises are intended to assist you in reviewing Chapter 10.

10.1 For the system discussed in Section 2, establish expressions for $\mathbf{H}_O$ when the disk of mass m_2 is replaced by a particle of mass m_2 which is attached at A. Is it possible to replace the system of the rigid rod and particle by a system consisting of a single rigid body?

10.2 For the system discussed in Exercise 10.1, derive an expression for the kinetic energy T.

10.3 Consider the system of rigid bodies discussed in Section 2. How do the results for $\mathbf{H}_O$ and T simplify if the disk were pin-jointed at its center of mass to A?

10.4 For the system discussed in Exercise 10.4, derive expressions for $\mathbf{H}_O$ and T if the disk were welded at its center of mass to A.

10.5 For the system of two rigid bodies discussed in Section 2, derive expressions for $\mathbf{H}_O$ and T if the disk were replaced by a rigid rod of length $2R$.

10.6 A circular disk of mass m_1 and radius R lies at rest on a horizontal plane. The origin of the coordinate system is taken to coincide with the center of mass of the disk. At an instant in time t_1, a particle of mass m_2 which has a velocity vector $\mathbf{v} = v_x \mathbf{E}_x + v_y \mathbf{E}_y$ collides with the disk. The collision occurs at the point of the disk whose position vector is $R \cos(\phi)\mathbf{E}_x + R \sin(\phi)\mathbf{E}_y$. After the impact, the particle adheres to the disk. Show that the position vector of the center of mass of the system during the instant of impact is

$$\bar{\mathbf{x}} = \frac{m_2 R}{m_1 + m_2}(\cos(\phi)\mathbf{E}_x + \sin(\phi)\mathbf{E}_y).$$

In addition, show that the velocity vector of the center of mass of the system immediately following the impact is

$$\bar{\mathbf{v}}(t_{1+}) = \frac{m_2}{m_1 + m_2}(v_x \mathbf{E}_x + v_y \mathbf{E}_y).$$

10.7 For the system discussed in Exercise 10.6, show that the angular momentum of the system relative to its center of mass at the instant prior to the collision is

$$\mathbf{H}(t_{1-}) = \frac{m_1 m_2 R}{m_1 + m_2}(v_{1x} \sin(\phi) - v_{1y} \cos(\phi))\mathbf{E}_z.$$

In addition, show that the angular momentum of the system relative to its center of mass immediately after the collision is

$$\mathbf{H}(t_{1+}) = \left(\frac{m_2}{m_1 + m_2} + \frac{1}{2}\right) m_1 R^2 \omega(t_{1+})\mathbf{E}_z,$$

where $\omega(t_{1+})\mathbf{E}_z$ is the angular velocity vector of the system.

10.8 Using the results of Exercise 10.7, determine the angular velocity vector of the system immediately following the impact discussed in Exercise 10.6. Under which conditions is it possible for this velocity vector to be $\mathbf{0}$?

10.9 Determine the kinetic energy lost during the collision discussed in Exercise 10.6.

10.10 Which modifications to the results of Exercises 10.6 to 10.8 would be needed to accommodate the situation where the center of mass of the disk was in motion at the instant prior to impact?

10.11 Repeat Exercises 10.6 through 10.9 for the case where the disk is pinned at its center to a fixed point O.

Appendix A
Preliminaries on Vectors and Calculus

Caveat Lector

In writing this primer, I have assumed that the reader has had courses in linear algebra and calculus. This being so, I have more often than not found that these topics have been forgotten. Here, I review some of the basics. But it is a terse review, and I strongly recommend that readers review their own class notes and other texts on these topics in order to fill the gaps in their knowledge.

Students who are able to differentiate vectors, and are familiar with the chain and product rules of calculus, have a distinct advantage in comprehending the material in this primer and in other courses. I have never been able to sufficiently emphasize this point to students at the beginning of an undergraduate dynamics course.

A.1 Vector Notation

A fixed (right-handed) Cartesian basis for Euclidean three-space $\mathcal{E}^3$ is denoted by the set $\{\mathbf{E}_x, \mathbf{E}_y, \mathbf{E}_z\}$. These three vectors are orthonormal (i.e., they each have a unit magnitude and are mutually perpendicular).

For any vector $\mathbf{b}$, one has the representation

$$\mathbf{b} = b_x \mathbf{E}_x + b_y \mathbf{E}_y + b_z \mathbf{E}_z\,,$$

where b_x, b_y, and b_z are the Cartesian components of the vector $\mathbf{b}$ (cf. Figure A.1).

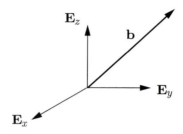

FIGURE A.1. A vector **b**

A.2   Dot and Cross Products

The two most commonly used vector products are the dot and cross products. The dot product of any two vectors **u** and **w** is a scalar defined by

$$\mathbf{u} \cdot \mathbf{w} = u_x w_x + u_y w_y + u_z w_z = \|\mathbf{u}\| \|\mathbf{w}\| \cos(\gamma),$$

where γ is the angle subtended by **u** and **w**, and $\|\mathbf{b}\|$ denotes the norm (or magnitude) of a vector **b**:

$$\|\mathbf{b}\|^2 = \mathbf{b} \cdot \mathbf{b} = b_x^2 + b_y^2 + b_z^2.$$

Clearly, if two vectors are perpendicular to each other, then their dot product is zero.

One can use the dot product to define a unit vector **n** in the direction of any vector **b**:

$$\mathbf{n} = \frac{\mathbf{b}}{\|\mathbf{b}\|}.$$

This formula is very useful in establishing expressions for friction forces and normal forces.

The cross product of any two vectors **b** and **c** is a vector that is perpendicular to the plane they define:

$$\mathbf{b} \times \mathbf{c} = (b_y c_z - b_z c_y)\,\mathbf{E}_x + (b_z c_x - b_x c_z)\,\mathbf{E}_y + (b_x c_y - b_y c_x)\,\mathbf{E}_z.$$

This expression for the cross product can be expressed in another form involving the determinant of a matrix:

$$\mathbf{b} \times \mathbf{c} = -\mathbf{c} \times \mathbf{b} = \det \begin{bmatrix} \mathbf{E}_x & \mathbf{E}_y & \mathbf{E}_z \\ b_x & b_y & b_z \\ c_x & c_y & c_z \end{bmatrix}.$$

You should notice that if two vectors are parallel, then their cross product is the zero vector **0**.

A.3   Differentiation of Vectors

Given a vector $\mathbf{u}$, suppose it is a function of time t: $\mathbf{u} = \mathbf{u}(t)$. One can evaluate its derivative using the product rule:

$$\frac{d\mathbf{u}}{dt} = \frac{du_x}{dt}\mathbf{E}_x + \frac{du_y}{dt}\mathbf{E}_y + \frac{du_z}{dt}\mathbf{E}_z + u_x\frac{d\mathbf{E}_x}{dt} + u_y\frac{d\mathbf{E}_y}{dt} + u_z\frac{d\mathbf{E}_z}{dt} \, .$$

However, $\mathbf{E}_x$, $\mathbf{E}_y$, and $\mathbf{E}_z$ are constant vectors (i.e., they have constant magnitude and direction). Hence, their time derivatives are zero:

$$\frac{d\mathbf{u}}{dt} = \frac{du_x}{dt}\mathbf{E}_x + \frac{du_y}{dt}\mathbf{E}_y + \frac{du_z}{dt}\mathbf{E}_z \, .$$

We can also use the product rule of calculus to show that

$$\frac{d}{dt}(\mathbf{u} \cdot \mathbf{w}) = \frac{d\mathbf{u}}{dt} \cdot \mathbf{w} + \mathbf{u} \cdot \frac{d\mathbf{w}}{dt} \, , \quad \frac{d}{dt}(\mathbf{u} \times \mathbf{w}) = \frac{d\mathbf{u}}{dt} \times \mathbf{w} + \mathbf{u} \times \frac{d\mathbf{w}}{dt} \, .$$

These results are obtained by representing the vectors $\mathbf{u}$ and $\mathbf{w}$ with respect to a Cartesian basis, then evaluating the left- and right-hand sides of both equations and showing their equality.

To differentiate any vector-valued function $\mathbf{c}(s(t))$ with respect to t, we use the chain rule:

$$\frac{d\mathbf{c}}{dt} = \frac{ds}{dt}\frac{d\mathbf{c}}{ds} = \frac{ds}{dt}\left(\frac{dc_x}{ds}\mathbf{E}_x + \frac{dc_y}{ds}\mathbf{E}_y + \frac{dc_z}{ds}\mathbf{E}_z\right) \, .$$

When differentiating it is important to distinguish a function and its value. For example, suppose $s = t^2$ and the function f as a function of time t is $f(t) = t$. Then, f as a function of s is $\hat{f}(s) = s$, but $f(s) = s^2 \neq \hat{f}(s)$.

A.4   A Ubiquitous Example of Vector Differentiation

One of the main sets of vectors arising in any course on dynamics is $\{\mathbf{e}_r, \mathbf{e}_\theta, \mathbf{e}_z\}$:

$$\begin{aligned}
\mathbf{e}_r &= \cos(\theta)\mathbf{E}_x + \sin(\theta)\mathbf{E}_y \, , \\
\mathbf{e}_\theta &= -\sin(\theta)\mathbf{E}_x + \cos(\theta)\mathbf{E}_y \, , \\
\mathbf{e}_z &= \mathbf{E}_z \, .
\end{aligned}$$

We also refer the reader to Figure A.2. In the above equations, θ is a function of time.

Using the previous developments, you should be able to show the following results:

$$\frac{d\mathbf{e}_r}{d\theta} = -\sin(\theta)\mathbf{E}_x + \cos(\theta)\mathbf{E}_y = \mathbf{e}_\theta \, ,$$

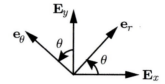

FIGURE A.2. The vectors $\mathbf{e}_r$ and $\mathbf{e}_\theta$

$$\frac{d\mathbf{e}_\theta}{d\theta} = -\cos(\theta)\mathbf{E}_x - \sin(\theta)\mathbf{E}_y = -\mathbf{e}_r \,,$$

$$\frac{d\mathbf{e}_r}{dt} = \frac{d\theta}{dt}\mathbf{e}_\theta \,, \quad \frac{d\mathbf{e}_\theta}{dt} = -\frac{d\theta}{dt}\mathbf{e}_r \,.$$

A useful exercise is to evaluate these expressions and graphically represent them for a given $\theta(t)$. For example, $\theta(t) = 10t^2 + 15t$.

Finally, you should be able to show that

$$
\begin{aligned}
\mathbf{e}_r \times \mathbf{e}_\theta = \mathbf{e}_z \,, \quad & \mathbf{e}_z \times \mathbf{e}_r &=& \quad \mathbf{e}_\theta \,, \quad & \mathbf{e}_\theta \times \mathbf{e}_z = \mathbf{e}_r \,, \\
\mathbf{e}_r \cdot \mathbf{e}_r = 1 \,, \quad & \mathbf{e}_\theta \cdot \mathbf{e}_\theta &=& \quad 1 \,, \quad & \mathbf{e}_z \cdot \mathbf{e}_z = 1 \,, \\
\mathbf{e}_r \cdot \mathbf{e}_\theta = 0 \,, \quad & \mathbf{e}_\theta \cdot \mathbf{e}_z &=& \quad 0 \,, \quad & \mathbf{e}_r \cdot \mathbf{e}_z = 0 \,.
\end{aligned}
$$

In other words, $\{\mathbf{e}_r, \mathbf{e}_\theta, \mathbf{e}_z\}$ forms an orthonormal set of vectors. Furthermore, since $\mathbf{e}_z \cdot (\mathbf{e}_r \times \mathbf{e}_\theta) = 1$, this set of vectors is also right-handed.

A.5   Ordinary Differential Equations

The main types of differential equations appearing in undergraduate dynamics courses are of the form $\ddot{u} = f(u)$, where the superposed double dot indicates the second derivative of u with respect to t. The general solution of this differential equation involves two constants: the initial conditions for $u(t_0) = u_0$, and its velocity $\dot{u}(t_0) = \dot{u}_0$. Often, one chooses time such that $t_0 = 0$.

The most comprehensive source of mechanics problems that involve differential equations of the form $\ddot{u} = f(u)$ is Whittaker's classical work [67]. It should also be added that classical works in dynamics placed tremendous emphasis on obtaining analytical solutions to such equations. Recently, the engineering dynamics community has become increasingly aware of possible chaotic solutions. Consequently, the existence of analytical solutions is generally not anticipated. We refer the reader to Moon [40] and Strogatz [62] for further discussions on, and references to, this matter. Further perspectives can be gained by reading the books by Barrow-Green [3] and Diacu and Holmes [20] on Henri Poincaré's seminal work on chaos, and Peterson's book [48] on chaos in the solar system.

A.5.1   The Planar Pendulum

One example of the above differential equation arises in the planar pendulum discussed in Chapter 2. Recall that the equation governing the motion of the pendulum was

$$mL\ddot{\theta} = -mg\cos(\theta)\,.$$

This equation is of the form discussed above with $u = \theta$ and $f(u) = -g\cos(u)/L$. Here, f is a nonlinear function of u. Given the initial conditions $\theta(t_0) = \theta_0$ and $\dot{\theta}(t_0) = \dot{\theta}_0$, this differential equation can be solved analytically. The resulting solution involves special functions that are known as Jacobi's elliptic functions.[1] Alas, these functions are beyond the scope of an undergraduate dynamics class, so instead one normally is required to use the conservation of the total energy E of the particle to solve most posed problems involving this pendulum.

A.5.2   The Projectile Problem

A far easier example arises in the motion of a particle under the influence of a gravitational force $-mg\mathbf{E}_y$. There, the differential equations governing the motion of the particle are

$$m\ddot{x} = 0\,, \quad m\ddot{y} = -mg\,, \quad m\ddot{z} = 0\,.$$

Clearly, each of these three equations is of the form $\ddot{u} = f(u)$. The general solution to the second of these equations is

$$y(t) = y_0 + \dot{y}_0(t - t_0) - \frac{g}{2}(t - t_0)^2\,.$$

Here, $y(t_0) = y_0$ and $\dot{y}(t_0) = \dot{y}_0$ are the initial conditions. You should verify the solution for $y(t)$ given above by first examining whether it satisfies the initial conditions and then seeing whether it satisfies the differential equation $\ddot{y} = -g$. By setting $g = 0$ and changing variables from y to x and z, the solutions to the other two differential equations can be obtained.

A.5.3   The Harmonic Oscillator

The most common example of a differential equation in mechanical engineering is found from the harmonic oscillator. Here, a particle of mass m is attached by a linear spring of stiffness K to a fixed point. The variable x is chosen to measure both the displacement of the particle and the displacement of the spring from its unstretched state. The governing differential equation is

$$m\ddot{x} = -Kx\,.$$

---

[1] A discussion of these functions, in addition to the analytical solution of the particle's motion, can be found in Lawden [36], for instance.

This equation has the general solution

$$x(t) = x_0 \cos\left(\sqrt{\frac{K}{m}}\,(t - t_0)\right) + \dot{x}_0 \sqrt{\frac{m}{K}} \sin\left(\sqrt{\frac{K}{m}}\,(t - t_0)\right),$$

where $x(t_0) = x_0$ and $\dot{x}(t_0) = \dot{x}_0$ are the initial conditions.

A.5.4   A Particle in a Whirling Tube

The last example of interest arises in problems concerning a particle of mass m that is in motion in a smooth frictionless tube. The tube is being rotated in a horizontal plane with a constant angular speed Ω. The differential equation governing the radial motion of the particle is

$$m\ddot{r} = m\Omega^2 r\,.$$

This equation has the general solution

$$r(t) = r_0 \cosh\left(\Omega\,(t - t_0)\right) + \frac{\dot{r}_0}{\Omega} \sinh\left(\Omega\,(t - t_0)\right),$$

where $r(t_0) = r_0$ and $\dot{r}(t_0) = \dot{r}_0$ are the initial conditions.

Appendix B
Weekly Course Content and Notation in Other Texts

Abbreviations

For convenience in this appendix, we shall use the following abbreviations: BF, Bedford and Fowler [6]; BJ, Beer and Johnston [7]; H, Hibbeler [33]; MK, Meriam and Kraige [39]; RS, Riley and Sturges [50]; and S, Shames [56].

B.1 Weekly Course Content

The following is an outline for a 15-week (semester-long) course in undergraduate engineering dynamics. Here, we list the weekly topics along with the corresponding sections in this primer. We also indicate the corresponding sections in other texts. This correspondence is, of course, approximate: all of the cited texts have differences in scope and emphasis.

Normally, the course is divided into three parts: a single particle, systems of particles, and (planar dynamics of) rigid bodies. The developments in most texts also cover the material in this order, the exception being Riley and Sturges [50].

Week Number	Topic	Primer Section	Other Texts
1	Single Particle: Cartesian Coordinates	Chapter 1	BF: Ch.1, 2.1–2.3, 3.1–3.4 BJ: 11.1–11.11, 12.5 H: 12.1–12.6, 13.4 MK: 1/1–1/7, 2/2, 2/4, 3/4 RS: 13.1–13.4, 15.1–15.3 S: 11.1–11.4, 12.1–12.4
2	Single Particle: Polar Coordinates	Chapter 2	BF: 2.3, 3.4 BJ: 11.14, 12.8 H: 12.8, 13.6 MK: 2/6, 3/5 RS: 13.5, 13.7, 15.4 S: 11.6, 12.5
3	Single Particle: Serret-Frenet Triads	Chapter 3	BF: 2.3, 3.4 BJ: 11.13, 12.5 H: 12.7, 13.5 MK: 2/5, 2/7, 3/5 RS: 13.5, 13.7, 15.4 S: 11.5, 12.9
4	Single Particle: Further Kinetics	Chapter 4	BF: 3.4 BJ: 12.5 H: 13.4–13.6 MK: 3/5 RS: 15.3, 15.4 S: 12.4, 12.5, 12.9
5	Single Particle Work and Energy	Chapter 5	BF: Ch. 4 BJ: 13.1–13.9 H: 14.1, 14.2, 14.4–14.6 MK: 3/6, 3/7 RS: 17.1–17.3, 17.5–17.10 S: 13.1–13.5
6	Single Particle: Linear and Angular Momentum	Chapter 6 Sects. 1 & 2	BF: 5.1, 5.2, 5.4 BJ: 12.2, 12.7, 12.9,13.11 H: 15.1, 15.5–15.7 MK: 3/9, 3/10 RS: 19.2, 19.5 S: 14.1, 14.3, 14.6

Week Number	Topic	Primer Section	Other Texts
7	Collisions of Particles	Chapter 6 Sects. 3–5	BF: 5.3 BJ: 13.12–13.14 H: 14.3, 14.6, 15.4 MK: 3/12 RS: 19.4 S: 14.4–14.5
8	Systems of Particles	Chapter 7	BF: 7.1, 8.1 BJ: 14.1–14.9 H: 13.3, 14.3, 14.6, 15.3 MK: 4/1–4/5 RS: 17.4–17.8, 19.3, 19.5 S: 12.10, 14.2, 14.7, 13.6–13.9
9	Kinematics of Rigid Bodies	Chapter 8	BF: 6.1–6.3 BJ: 15.1–15.4 H: 16.1–16.4 MK: 5/1–5/4 RS: 14.1–14.3 S: 15.1–15.5
10	Kinematics of Rigid Bodies	Chapter 8	BF: 6.4–6.6 BJ: 15.4–15.8, 15.10–15.15 H: 16.4–16.8 MK: 5/5–5/7 RS: 14.4–14.6 S: 15.5–15.11
11	Planar Dynamics of Rigid Bodies	Chapter 9	BF: 7.2–7.3, App., 9.2 BJ: Ch. 16 H: 21.1, 21.2, 17.3 MK: 6/1–6/3, Apps. A & B RS: 16.2, 16.3, 20.6 S: 16.1–16.4
12	Planar Dynamics of Rigid Bodies	Chapter 9	BF: 7.4 BJ: Ch. 16 H: 17.4 MK: 6/4 RS: 16.4 S: 16.5

Week Number	Topic	Primer Section	Other Texts
13	Planar Dynamics of Rigid Bodies	Chapter 9	BF: 8.1–8.3 BJ: Ch. 16 & 17.1–17.7 H: 17.5 & Ch. 18 MK: 6/5. 6/6 RS: 16.4 & Ch. 18 S: 16.6, 17.1–17.3
14	Planar Dynamics of Rigid Bodies	Chapter 10	BF: 8.4 BJ: 17.8–17.11 H: Ch. 19 MK: 6/8 RS: 20.1–20.5 S: 17.4–17.7
15	Vibrations	Not Covered	BF: Ch. 10 BJ: Ch. 19 H: Ch. 2 MK: Ch. 8 RS: Ch. 21 S: Ch. 22

B.2   Notation in Other Texts

Here, we give a brief summary of some of the notational differences between this primer and those used in other texts. In many of the cited texts only plane curves are considered. Consequently, the binormal vector $\mathbf{e}_b$ is not explicitly mentioned.

	Primer Notation	Other Texts
Cartesian Basis Vectors	$\{\mathbf{E}_x, \mathbf{E}_y, \mathbf{E}_z\}$	BF: $\{\mathbf{i}, \mathbf{j}, \mathbf{k}\}$ BJ: $\{\mathbf{i}, \mathbf{j}, \mathbf{k}\}$ H: $\{\mathbf{i}, \mathbf{j}, \mathbf{k}\}$ MK: $\{\mathbf{i}, \mathbf{j}, \mathbf{k}\}$ RS: $\{\mathbf{i}, \mathbf{j}, \mathbf{k}\}$ S: $\{\mathbf{i}, \mathbf{j}, \mathbf{k}\}$
Serret-Frenet Triad	$\{\mathbf{e}_t, \mathbf{e}_n, \mathbf{e}_b\}$	BF: $\{\mathbf{e}_t, \mathbf{e}_n, -\}$ BJ: $\{\mathbf{e}_t, \mathbf{e}_n, \mathbf{e}_b\}$ H: $\{\mathbf{u}_t, \mathbf{u}_n, \mathbf{u}_b\}$ MK: $\{\mathbf{e}_t, \mathbf{e}_n, -\}$ RS: $\{\mathbf{e}_t, \mathbf{e}_n, -\}$ S: $\{\boldsymbol{\epsilon}_t, \boldsymbol{\epsilon}_n, \boldsymbol{\epsilon}_t \times \boldsymbol{\epsilon}_n\}$
Linear Momentum of a Particle	$\mathbf{G} = m\mathbf{v}$	BF: $m\mathbf{v}$ BJ: $\mathbf{L} = m\mathbf{v}$ H: $m\mathbf{v}$ MK: $\mathbf{G}$ RS: $\mathbf{L} = m\mathbf{v}$ S: $m\mathbf{V}$
Corotational Basis or Body Fixed Basis	$\{\mathbf{e}_x, \mathbf{e}_y, \mathbf{e}_z\}$	BF: $\{\mathbf{i}, \mathbf{j}, \mathbf{k}\}$ BJ: $\{\mathbf{i}, \mathbf{j}, \mathbf{k}\}$ H: $\{\mathbf{i}, \mathbf{j}, \mathbf{k}\}$ MK: $\{\mathbf{i}, \mathbf{j}, \mathbf{k}\}$ RS: $\{\mathbf{e}_x, \mathbf{e}_y, \mathbf{e}_z\}$ S: $\{\mathbf{i}, \mathbf{j}, \mathbf{k}\}$

References

[1] S. S. Antman, *Nonlinear Problems of Elasticity*, Springer-Verlag, New York (1995).

[2] V. I. Arnol'd, *Mathematical Methods in Classical Mechanics*, Springer-Verlag, New York (1978).

[3] J. Barrow-Green, *Poincaré and the Three-Body Problem*, American Mathematical Society, Providence (1997).

[4] M. F. Beatty, "Kinematics of finite rigid body displacements," *American Journal of Physics*, **34**, pp. 949–954 (1966).

[5] M. F. Beatty, *Principles of Engineering Mechanics, I. Kinematics – The Geometry of Motion*, Plenum Press, New York (1986).

[6] A. Bedford and W. Fowler, *Engineering Mechanics - Dynamics*, Addison-Wesley, Reading, Massachusetts (1995).

[7] F. P. Beer and E. R. Johnston, Jr., *Vector Mechanics for Engineers: Dynamics*, Fifth Edition, McGraw-Hill, New York (1988).

[8] R. Bellman, *Introduction to Matrix Analysis*, McGraw-Hill, New York (1960).

[9] O. Bottema and B. Roth, *Theoretical Kinematics*, North-Holland, New York (1979).

[10] R. M. Brach, *Mechanical Impact Dynamics: Rigid Body Collisions*, John Wiley & Sons, New York (1991).

[11] J. Casey, "A treatment of rigid body dynamics," *ASME Journal of Applied Mechanics*, **50**, pp. 905–907 (1983) and **51**, p. 227 (1984)

[12] J. Casey, *Elements of Dynamics*, Unpublished Manuscript, Department of Mechanical Engineering, University of California at Berkeley (1993).

[13] J. Casey, "Geometrical derivation of Lagrange's equations for a system of particles," *American Journal of Physics*, **62**, pp. 836–847 (1994).

[14] J. Casey, "On the advantages of a geometrical viewpoint in the derivation of Lagrange's equations for a rigid continuum," *Journal of Applied Mathematics and Physics (ZAMP)*, **46** (Special Issue) pp. S805–S847 (1995).

[15] G. Coriolis, *Théorie Mathématique des effets du Jeu de Billard*,[1] Carilian-Gouery, Paris (1835).

[16] C. A. Coulomb, "Théorie des machines simples en ayant égard au frottement et à la roideur des cordages"[2] *Mémoires de Mathématique et de Physique présentés à l'Académie Royale des Sciences par divers Savans, et lus dans ses assemblés*, **10**, pp. 161–332 (1785).

[17] H. Crabtree, *An Elementary Treatment of the Theory of Spinning Tops and Gyroscopes*, Third Edition, Chelsea Publishing, New York (1967).

[18] R. Cushman, J. Hermans, and D. Kemppainen, "The rolling disk," in *Nonlinear Dynamical Systems and Chaos*, edited by H. W. Broer, S. A. van Gils, I. Hoveijn, and F. Takens, *Progress in Nonlinear Differential Equations and their Applications*, **19**, pp. 21–60. Birkhäuser, Basel (1996).

[19] G. Darboux, *Leçons sur la Théorie Générale des Surfaces et les Applications Géométriques du Calcul Infinitesimal*,[3] Parts 1–4, Gauthier-Villars, Paris (1887–1896).

[20] F. Diacu and P. Holmes, *Celestial Encounters: The Origins of Chaos and Stability*, Princeton University Press, Princeton (1996).

---

[1] The title of this book translates to *Mathematical Theory of Effects on the Game of Billiards*.

[2] The title of this paper translates to "Theory of simple machines with consideration of the friction and rubbing of rigging."

[3] The title of this book translates to *Lectures on the General Theory of Surfaces and Geometric Applications of Calculus*.

[21] R. Dugas, *A History of Mechanics*, (translated from French by J. R. Maddox), Dover Publications, New York (1988).

[22] L. Euler, "Recherches sur le mouvement des corps célestes en général,"[4] *Mémoires de l'Académie des Sciences de Berlin*, **3**, pp. 93–143 (1749). Reprinted in *Leonhardi Euleri Opera Omnia*, Series Secunda, **25**, pp. 1–44, edited by M. Schürer, Zürich, Orell Füssli (1960).

[23] L. Euler, "Decouverte d'un nouveau principe de mechanique,"[5] *Mémoires de l'Académie des Sciences de Berlin*, **6**, pp. 185–217 (1752). Reprinted in *Leonhardi Euleri Opera Omnia*, Series Secunda, **5**, pp. 81–108, edited by J. O. Fleckenstein, Zürich, Orell Füssli (1957).

[24] L. Euler, "Nova methodus motum corporum rigidorum determinandi,"[6] *Nova Commentarii Academiae Scientiarum Petropolitanae*, **20**, pp. 208–238 (1776). Reprinted in *Leonhardi Euleri Opera Omnia*, Series Secunda, **9**, pp. 99–125, edited by C. Blanc, Zürich, Orell Füssli (1968).

[25] M. Fecko, "Falling cat connections and the momentum map," *Journal of Mathematical Physics*, **36**, pp. 6709–6719 (1995).

[26] J.-F. Frenet, "Sur quelques propriétés des courbes à double courbure,"[7] *Journal de Mathématiques pures et appliquées*, **17**, pp. 437–447 (1852).

[27] T. D. Gillespie, *Fundamentals of Vehicle Dynamics*, Society of Automotive Engineers, Warrendale (1992).

[28] W. Goldsmith, *Impact: The Theory and Physical Behavior of Colliding Solids*, Arnold, New York (1960).

[29] D. T. Greenwood, *Principles of Dynamics*, Second Edition, Prentice-Hall, Englewood Cliffs, New Jersey (1988).

[30] M. E. Gurtin, *An Introduction to Continuum Mechanics*, Academic Press, San Diego (1981).

[31] J. Hermans, "A symmetric sphere rolling on a surface," *Nonlinearity*, **8**, pp. 493–515 (1995).

---

[4]The title of this paper translates to "Researches on the motion of celestial bodies in general."

[5]The title of this paper translates to "Discovery of a new principle of mechanics."

[6]The title of this paper transltates to "A new method to determine the motion of rigid bodies."

[7]The title of this paper translates to "On several properties of curves of double curvature." According to Spivak [59], curves of double curvature is an old term for space curves.

[32] J. Heyman, *Coulomb's Memoir on Statics: An Essay in the History of Civil Engineering*, Cambridge University Press, Cambridge (1972).

[33] R. C. Hibbeler, *Engineering Mechanics*, Eighth Edition, Prentice-Hall, Upple Saddle River, New Jersey (1997).

[34] T. R. Kane and M. P. Scher, "A dynamical explanation of the falling cat phenomenon," *International Journal of Solids and Structures*, **5**, pp. 663–670 (1969).

[35] E. Kreyszig, *Differential Geometry*, Revised Edition, University of Toronto Press, Toronto (1964).

[36] D. F. Lawden, *Elliptic Functions and Applications*, Springer-Verlag, New York (1989).

[37] A. E. H. Love, *A Treatise on the Mathematical Theory of Elasticity*, Fourth Edition, Cambridge University Press, Cambridge (1927).

[38] H. H. Mabie and F. W. Ocvirk, *Mechanisms and Dynamics of Machinery*, Third Edition, John Wiley & Sons, New York (1978).

[39] J. L. Meriam and L. G. Kraige, *Engineering Mechanics: Dynamics*, Fourth Edition, John Wiley & Sons, New York (1997).

[40] F. C. Moon, *Chaotic and Fractal Dynamics: An Introduction for Applied Scientists and Engineers*, John Wiley & Sons, New York (1992).

[41] F. R. Moulton, *An Introduction to Celestial Mechanics*, Second Edition, Macmillan, New York (1914).

[42] Ju. I. Neimark and N. A. Fufaev, *Dynamics of Nonholonomic Systems*, translated from Russian by J. R. Barbour, American Mathematical Society. Providence, Rhode Island (1972).

[43] I. Newton, *Philosophiae Naturalis Principia Mathematica*. Originally published in London in 1687. English translation in 1729 by A. Motte, revised translation by F. Cajori published by University of California Press, Berkeley (1934).

[44] I. Newton, *The Mathematical Papers of Isaac Newton*, **5**, edited by D. T. Whiteside, Cambridge University Press, Cambridge (1972).

[45] O. M. O'Reilly, "On the dynamics of rolling disks and sliding disks," *Nonlinear Dynamics*, **10**, pp. 287–305 (1996).

[46] P. Painlevé, "Sur les lois du frottement de glissement,"[8] *Comptes Rendus Hebdomadaires des Séances de l'Académie des Sciences*, **121**, pp.

---

[8]The title of this paper translates to "On the laws of sliding friction."

112–115 (1895), **140**, 702–707 (1905), **141**, 401–405 (1905), and **141**, 546–552 (1905).

[47] B. Paul, *Kinematics and Dynamics of Planar Machinery*, Prentice-Hall, Englewood Cliffs, New Jersey (1979).

[48] I. Peterson, *Newton's Clock: Chaos in the Solar System*, W. H. Freeman and Company, New York (1993).

[49] E. Rabinowicz, *Friction and Wear of Materials*, Second Edition, John Wiley & Sons, New York (1995).

[50] W. F. Riley and L. D. Sturges, *Engineering Mechanics: Dynamics*, Second Edition, John Wiley & Sons, New York (1996).

[51] E. J. Routh, *The Elementary Part of a Treatise on the Dynamics of a System of Rigid Bodies*, Seventh Edition, Macmillan, London (1905).

[52] E. J. Routh, *The Advanced Part of a Treatise on the Dynamics of a System of Rigid Bodies*, Sixth Edition, Macmillan, London (1905).

[53] M. B. Rubin, "Physical restrictions on the impulse acting during three-dimensional impact of two rigid bodies," *ASME Journal of Applied Mechanics*, **65**, pp. 464–469 (1998).

[54] A. Ruina, "Constitutive relations for frictional slip," in *Mechanics of Geomaterials: Rocks, Concrete, Soils*, edited by Z. P. Bažant, pp. 169–199, John Wiley & Sons, New York (1985).

[55] J. A. Serret, "Sur quelques formules relatives à la théorie des courbes à double courbure,"[9] *Journal de Mathématiques pures et appliquées*, **16**, pp. 193–207 (1851).

[56] I. H. Shames, *Engineering Mechanics – Statics and Dynamics*, Fourth Edition, Prentice-Hall, Upper Saddle River, New Jersey (1997).

[57] A. Shapere and F. Wilczek, "Gauge kinematics of deformable bodies," *American Journal of Physics*, **57**, pp. 514–518 (1989).

[58] M. D. Shuster, "A survey of attitude representations," *The Journal of the Astronautical Sciences*, **41**, pp. 439–517 (1993).

[59] M. Spivak, *A Comprehensive Introduction to Differential Geometry*, **2**, Second Edition, Publish or Perish, Berkeley (1979).

---

[9]The title of this paper translates to "On several formulae relating to curves of double curvature." According to Spivak [59], curves of double curvature is an old term for space curves.

[60] D. E. Stewart, "Rigid-body dynamics with friction and impact," *SIAM Review*, **42**, pp. 3–39 (2000).

[61] G. Strang, *Linear Algebra and its Applications*, Third Edition, Harcourt Brace Jovanovich Publications, San Diego (1988).

[62] S. H. Strogatz, *Nonlinear Dynamics and Chaos, with Applications to Physics, Chemistry, Biology and Engineering*, Addison-Wesley, Reading (1994).

[63] D. J. Struik, *Lectures on Classical Differential Geometry*, Second Edition, Dover Publications, New York (1988).

[64] J. L. Synge and B. A. Griffith, *Principles of Mechanics*, McGraw-Hill, New York (1942).

[65] C. Truesdell, *Essays on the History of Mechanics*, Springer-Verlag, New York (1968).

[66] C. Truesdell and R. A. Toupin, *The Classical Field Theories*, in *Handbuch der Physik*, **3/1**, edited by S. Flügge, Springer-Verlag, Berlin (1960).

[67] E. T. Whittaker, *A Treatise on the Analytical Dynamics of Particles and Rigid Bodies*, Fourth Edition, Dover Publications, New York (1944).

[68] D. V. Zenkov, A. M. Bloch, and J. E. Marsden, "The energy-momentum method for the stability of non-holonomic systems," *Dynamics and Stability of Systems*, **13**, pp. 123–165 (1998).

Index